Tributes
Volume 27

Why is this a Proof?
Festschrift for Luiz Carlos Pereira

Tributes Series Editor
Dov Gabbay

dov.gabbay@kcl.ac.uk

Why is this a Proof?
Festschrift for Luiz Carlos Pereira

edited by

Edward Hermann Haeusler,
Wagner de Campos Sanz
and
Bruno Lopes

ISBN 978-1-84890-172-8

College Publications
Scientific Director: Dov Gabbay
Managing Director: Jane Spurr

http://www.collegepublications.co.uk

Cover design by Laraine Welch

Printed by Lightning Source, Milton Keynes, UK

Contents

Preface

From an academic lineage that remounts to the first Professor of Theoretical Philosophy of the University of Stockholm (Anders Wedberg), through one of the most distinguished logicians of our days (Dag Prawitz), Luiz Carlos Pereira presented to the world his work *"On the Estimation of the Length of Normal Derivations"* in 1982. So far he has published more than 65 works and keep spreading his knowledge with more than 25 former students.

This volume includes fifteen research papers with the purpose of celebrating Luiz Carlos Pereira's 60th birthday. Among the authors contributing to the volume we find colleagues, friends — including his PhD advisor — and admirers. Similar to Luiz Carlos Pereira's intelectual interests and work, the contributions range from Philosophy to Mathematics, from Mathematics to Logic, and from Logic to Philosophy again, passing through Computer Science. They are the result of current research by well-known scholars in these fields. Proof Theory is, maybe, the Ariadne's thread uniting the different subjects treated. Questions around the nature of proofs are often present in Luiz Carlos formal and informal talks and publications. Of course, he was not the first to dedicated himself to these questions, but he raises and deal with them quite enthusiastically always. This enthusiasm together with his intelectual perspicacity has made us to enjoy a wonderful journey into the world of proofs. Although not all contributions are directly on proof theory, we can feel its echoes in all of them.

Luiz Carlos Pereira is one of the most distinguished Brazilian logicians. His work regarding Logic in natural deduction, proof theory, normalization and constructivity have affected researchers from all of the world for at least four academic generations.

Luiz Carlos is not known only by his high-quality and cutting-edge work. Kindness, generosity and helpfulness are strong features ofs his personality. This book aims to be not only a celebration of his career but also a tribute to the great person that is Luiz Carlos.

Great researchers collaborated in this project. Lassalle-Casanave traces ideas in Leibniz and Hilbert works to better understand Hilbert's formalism. Dag Prawitz presents a discussion on classical *versus* intuitionisc logic focused on deductive reasoning. Motivated by researches supervised by Luiz Carlos, Sautter conducts two experiments with negation inspired by Syllogistics and by translations between logics. Gisele Secco revisits the Four Color Theorem discussing the philosophical controversies it bring forwards.

Focused on Wittgenstein's criticism of Cantor's diagonalization proof, Imaguire tries to make sense of some of his comments on transfinite numbers. Legris presents some remarks on proof-theoretic semantics, discussing its idea, usages of language and the universalist perspective. A proof that the ordered structure of the Dedekind real numbers object is homogeneous in any topos with natural numbers object is presented by Coniglio and Sbardellini.

Reconsidering topics discussed with Luiz Carlos, Oswaldo Chateaubriand discusses the problem of the denotation of sentences. For "most", "many", "several" and other "generally" quantifications, Veloso et al. examine the structure of natural deduction derivations. Schroeder-Heister sketches an approach to proof-theoretic validity based on elimination rules and assesses its merits and limitations. The consequences of adding a constant to intuitionistic logic are explored by Ertola as a sort of local switch to classical logic.

Valeria de Paiva presents first steps towards a categorical version of the Bounded Functional Interpretation. Providing a philosophical justification to paraconsistent logics, Carnielli and Rodrigues also outline an analysis of contradiction, negation and consistency. Finally, Gilles Dowek discusses the definition of the classical connectives and quantifiers in constructive logic, proposing a definition of meaning of a small set of primitive connectives and quantifiers to explicitly define the others.

Although these works represent a large body of fields, we do not have the pretension of pointing all the fields Luiz Carlos acted in only one book.

It is just a demonstration of some topics he influenced and how many people admire his carrear. This book may be seen as a thank you note for Luiz Carlos. Thank you for all the great works. Thank you for all these decades of collaboration. Thank you for all the decades of partnership that hopefully will came.

Edward Hermann Haeusler *(PUC-Rio, Brazil)*
Wagner Sanz *(UFG, Brazil)*
Bruno Lopes *(UFF, Brazil)*

Acknowledgements

We would like to thank all the authors who joined us in this project and Jane Spurr for all the help during the editorial process. We would like also to thank all the agencies that supported (partially or not) the works in here presented, namely CONICET (Argentina), FAPESP (Brazil), CNPq (Brazil), DFG (Germany), FP7 (Europe) and FAPEMIG (Brazil).

On some functions of symbols in Hilbert's formalism[‡]

Abel Lassalle Casanave*

* Universidade Federal da Bahia
abel.lassalle@gmail.com

1 Introduction

At the beginning of his *Mathematical Logic*, Shoenfield gives two reasons for studying axioms and theorems as sentences:

> The first is that if we choose the language for expressing the axioms suitably, then the structure of sentences will reflect to some extent the meaning of the axiom. Thus we can study the concepts of the axiom system by studying the structure of the sentences expressing the axioms. This is particularly valuable for modern axiom systems, since for them our initial understanding of the basic concepts may be very weak.
>
> The second reason is that the concepts of mathematics are usually very abstract and therefore difficult to understand. A sentence, on the other hand, is a concrete object; so by studying axioms as sentences, we approach the abstract through the concrete. (Shoenfield, 1967, p. 2)

We can find in these passages some ideas related to Leibnizean symbolic knowledge. First of all, the sensibilization of thought by symbols subrogating 'meanings'. But this subrogation is, according to Shoenfield, of a special kind: the composition of the formulas reflects their meanings. This function of exhibition by symbols was called by Leibniz *ecthetic*. In this compositional exhibition we find the possibility of a calculus in the broader sense of ruled manipulation of symbols. Thus, according to Leibniz, symbols also accomplish a calculus function. Of course, in this sense,

---

[‡]I would like to thank Oswaldo Chateaubriand (PUC-Rio / Brazil), Wagner Sanz (UFG / Brazil) and Javier Legris (UBA / Argentine) for their comments on a preliminary version of this paper.

a proof in a formal system F, or, briefly, a formal proof, is a calculus; in Leibnizian terms, the symbols perform a calculus function. From these functions follows a particular kind of cognitive certainty that Leibniz called *ante oculos* certainty: we can visually verify if the ruled manipulation of symbols has been correctly applied. What kind of knowledge is this in which we disregard meanings in favor of blind symbolic manipulation? According to Leibniz, this is blind or symbolic knowledge (*cognitio caeca* or *cognitio symbolica*). But, why do we need symbolic knowledge?

In his second reason for studying formal systems, Shoenfield observes that mathematical concepts are (usually) very abstract, while sentences are concrete objects. In this way, by studying sentences instead of meanings we approach the abstract through the concrete. But, by 'abstract concepts', Shoenfield means essentially 'infinitary or transfinite concepts'; thus, we approach the abstract through the concrete sentences, that is, the infinite through the finite sequences of symbols. Of course, this approach is possible because the sequences of symbols exhibit the corresponding meanings.

In a more philosophically elaborated account, Leibniz distinguishes intuitive thinking and knowledge from symbolic thinking and knowledge. When knowledge is obtained by direct and simultaneous consideration of the simplest notes that compose a distinct idea, the knowledge is intuitive. This characterization implies that intuitive knowledge is somewhat difficult to achieve. It can be said that it is merely an ideal kind of knowledge. In general terms, our knowledge (and our thought) is symbolic, namely, it is obtained by means of symbols or characters. We cannot have intuitive knowledge about transfinite concepts, but we can have symbolic knowledge.

According also to Leibniz, by means of symbols we can express the infinite through the finite; we approach, in Shoenfield's terms, the abstract through the concrete. In fact, for Leibniz, mathematical symbols also accomplish a function of finitization, a strong form of sensibilization of thought. But the mathematical knowledge obtained by symbolic means will be just suppositive knowledge unless we prove that the symbolic apparatus does not imply a contradiction. In modern parlance, we need a consistency proof of the corresponding formal system.

The passages quoted from Shoenfield can illustrate the survival of Leibniz's ideas in modern logic. But our purpose in this paper is to show that Hilbert's formalism is a very important chapter in the history of symbolic knowledge[1]. First, we will briefly characterize Hilbert's formalism as

---

[1]In Esquisabel (2012) we find part of this history with a detailed discussion of the calculus function, the ecthetic function and the psychotecnic function of symbols. For the finitization

methodological; second we will distinguish three modes of founding symbolic knowledge. We will also briefly identify finitary arithmetical knowledge as a succedaneum of intuitive knowledge. Finally, we will show that mathematical knowledge is for Hilbert essentially symbolic formal knowledge.

2 Hilbert's formalism

What kind of formalism is defended by Hilbert? In our opinion, firstly, Hilbert's formalism is a conception according to which mathematics is conceived as a formal system deprived of meaning only from the perspective of proof theory. The theories in mathematical practice are theories "with content (=contentual)", but certainly this content is not intuitive (reliable), in any technical sense of the word. The problem was to justify such non-intuitive (unreliable) content by means of a consistency proof. From the foundational point of view, of course, this non-intuitive content is no content at all. We characterize this kind of formalism as methodological[2].

Secondly, methodological formalism is supported by a very strong thesis, namely that mathematics can be completely formalized. The formal theory, or the corresponding formal system, includes the formalization of the underlying logic, which is not made explicit in mathematical practice. The problem of proving that a formal theory is consistent generates a new mathematical domain: meta-mathematics or proof theory, which is mathematics with intuitive content in a technical sense.

Thirdly, methodological formalism is founded in a conception of mathematical knowledge we find in the tradition of Leibnizian symbolic knowledge. Three modes of founding symbolic knowledge in Leibniz can be distinguished, namely, that of being succedaneum of intuitive knowledge; as an instrumental extension of intuitive knowledge; and as formal knowledge. The first mode is in principle equivalent to intuitive knowledge: we use signs in the place of "things or concepts" for the sake of economical thinking. In this case, Leibniz has an explanation of symbolic knowledge in terms of the similarity between the symbolic system and the things or concepts subrogated:

> There is some relation or order in the characters which is also in things, especially if the characters are well invented. (GP VII, 191-192; Loemker, 1969, p. 184)

---

function, see Lassalle Casanave (2010).

[2]We used this interpretation of Hilbert's formalism as an 'operative concept' in Lassalle Casanave (1995) and Lassalle Casanave (1996) and other papers on Hilbert's Program.

But, if we pretend to reach knowledge by accepting non-intuitive entities or non-intuitive concepts introduced by means of signs, we can think, on the one hand, that symbolic manipulation of these signs is an eliminable device to reach intuitive knowledge in an abbreviated manner. We can speak in this case of ideal entities or ideal concepts corresponding to this symbolic device. In a letter to Varignon, Leibniz writes:

> It follows from this that even if someone refuses to admit infinite and infinitesimal lines in a rigorous metaphysical sense and as real things, he can still use them with confidence as ideal concepts which shorten his reasoning, similar to what we call imaginary roots in the ordinary algebra, for example $\sqrt{-2}$.
> (GM IV, 91-95; Loemker, 1969, p. 543)

But, on the other hand, we can think that symbolic knowledge furnishes formal knowledge in the sense of knowledge of forms or relations. From the conceptual point of view, this is the most important foundation of symbolic knowledge. In *On Universal Synthesis and Analysis, or the Art of Discovery and Judgment*, Leibniz says:

> For the rest, the art of combinations in particular, as I take it (it can also be called a general characteristic or algebra), is that science in which are treated the forms or formulas of things in general, that is, quality in general or similarity and dissimilarity; in the same way that ever new formulas arise from the elements a, b, c themselves when combined with each other, whether these elements represent quantities or something else.
> (GP VII, 292-298; Loemker, 1969, p. 233)

According to this view, 'common algebra' can be seen as providing knowledge about quantitative relations, that is, symbolic knowledge as formal knowledge. Now, the *ecthetic* function of symbols consists in presenting forms or relations, but not in the subrogation of previously given things or concepts[3].

Hilbert's concept of formal theory expresses the idea of this kind of symbolic knowledge. In fact, what is a formal theory for Hilbert? In a letter to Frege, he writes:

> But it is surely obvious that every theory is only a scaffolding or schema of concepts together with their necessary relations to one another, and that the basic elements can be thought of in any way one likes (Frege, 1980, p. 42).

---

[3]In Lassalle Casanave (2012*a*) these modes of founding symbolic knowledge are presented.

In this passage we find the very idea of symbolic knowledge as formal knowledge.[4] But this concept of 'formal' is not a foundational one; it is not enough for the task of proof theory. In mathematical practice are mixed mathematical symbolism and natural language; however, from the foundational point of view, it is necessary to abandon the verbal apparatus in favor of a purely symbolic one. Thus, besides the usual mathematical symbolism, the expressions of a formal language include logical symbols too. In this way, mathematics becomes a stock of formulas from the perspective of meta-mathematics, as Hilbert describes in *The Grounding of Elementary Number Theory*:

> Certain formulae that serve as a foundation for the formal edifice of mathematics are called axioms. A proof is a figure, which must be intuitively presented to us as such; it consists of inferences, where each of the premises is either an axiom, or agrees with the end-formula of an inference that comes earlier in the proof, or results from such a formula by substitution. (Hilbert, 1931, p. 1152)

In the transition from the concept of formal system in *Foundations of geometry* to this meta-mathematical concept of formal system the logical calculus is essential. According to Hilbert, the logical calculus was developed in a different context: its signs were introduced with the purpose of communication. But, in the meta-mathematical context, the formulas of the logical calculus do not mean anything in themselves. In the logical calculus we have a language capable of expressing mathematical propositions in formulas and also of expressing contentual inference as a formal process. The transition to this kind of formal treatment is accomplished:

> [O]n the one hand, for the axioms themselves, which originally were naively taken to be fundamental truths but in modern axiomatics had already for a long time been regarded as merely establishing certain interrelations between notions, and, on the other, for the logical calculus, which originally was to be only another language. (Hilbert, 1926, p. 381)

A formal theory in this merely syntactic sense represents the formal theory in practice; yet, it is considered deprived of content from the perspective of proof theory. That does not mean, of course, that Hilbert conceives mathematics as a mere "formula-game", although he certainly

---

[4]The same idea in Hibert (1918, p. 1108): The framework of concepts is nothing other than the theory of the field of knowledge.

stressed that symbolic manipulation is essential to mathematical knowledge. In the constructivist approach to mathematics, intuition is the only warrant for mathematical knowledge. But intuition implies always some kind of restriction to mathematically admissible concepts, principles or methods. Symbolic manipulations preserve these concepts, principles and methods.

Thus, the methodological formalism is grounded in an epistemological alternative to intuitive knowledge, namely, symbolic knowledge as formal knowledge. It is through a symbolic apparatus that *de facto* we use concepts, principles and methods which are not accessible to intuition. The notion of formal theory as a stock of formulas can be seen as the natural conclusion of a conception which recognizes the role of symbolism as an alternative to a limited intuition. And it is through a successful consistency proof that theories containing suspicious concepts or principles or methods could be *de iure* legitimated.

3   Some functions of symbols in Hilbert's Program

The (non-axiomatic) finitary number theory describes the domain of the intuitively acceptable.[5] In finitary arithmetic, variables subrogate numerical signs /, //, ///, ..., as well as other symbols subrogate finitary operations, finitary properties of them or finitary relations among them. One could say that numbers and number-concepts are, in this way, sensibilized by signs. But the logical operations of negation, unbounded universal quantification and unbounded existential quantification imply the transfinite totality of variable values.

However, the methods of the elementary theory of numbers already go beyond finitary arithmetic, as showed by the use of algebraic calculus:

> This theory [finitary arithmetic] always uses formulas for communication only; letters stand for numerals, and the fact that two signs are identical is communicated by an equation. In algebra, on the other hand, we consider the expressions formed with letters to be independent objects in themselves, and the contentual propositions of number theory are formalized by means of them. (Hilbert, 1926, p. 379)

---

[5]A formalization of the intuitive fragment of arithmetic can present a mathematical, not a foundational interest: its consistency follows from its intuitive character. The interest in the formalization *from the foundational point of view* consists precisely in the possibility of proving the consistency of suspicious concepts and methods.

Thus, by symbolic manipulation in finitary arithmetic we reach symbolic knowledge in the weak sense of a succedaneum of intuitive knowledge. But, how do we handle transfinite concepts, propositions or methods that we use in number theory? In an algebraic way: by blind manipulation of symbols according to rules. Thus, in a formal theory, transfinite concepts, propositions and methods are sensibilized by signs.

Let us consider *tertium non datur*. The combination of negation and unbounded quantification allows the formulation of this principle in its usual version in the predicate calculus, the touchstone of the dispute with the intuitionists. Hilbert writes:

> It is precisely one of the most important tasks of proof theory to clearly present the sense and admissibility of negation: negation is a formal process, by means of which, from a statement S, another arises, which is bound to S by the axioms of negation mentioned above (essentially, the principle of contradiction and *tertium non datur*). The process of negation is a necessary means of theoretical investigation; its unconditional application first makes possible the completeness and closure of Logic. But in general the statement that arises through negation is an ideal statement, and to take this ideal statement as being a real statement would be to misunderstand the nature and essence of thought. (Hilbert, 1931, pp. 1156–1157.)

From the foundational point of view *tertium non datur* has no content, it is an ideal statement; however, from the point of view of mathematical practice, it is a real, contentual but unreliable statement (methodological formalism!):

> The tertium non datur occupies a distinguished position among the axioms and theorems of logic in general: for while all the other axioms and theorems can be immediately traced back without difficulty to definitions, the tertium non datur expresses a new, contentually meaningful fact that stands in need of proof. (Hilbert, 1930, p. 125, *Apud* W. Ewald & W. Sieg 2013, p. 26)

In the above mentioned sensibilization of transfinite concepts by symbolic means the function of calculus in symbolic knowledge appears clearly. In fact, in place of contentual inference, eventually unreliable, "in proof theory we have an external action according to rules" (Hilbert, 1931, p. 1152). The external action according to rules is (logical) calculus.[6] And, of

---

[6]It is interesting to note the appearance in Hilbert's finitary arithmetic of what Poincaré

course, calculus guarantees *ante oculos* certainty. We cannot adopt without precautions the usual and problematic modes of inference, not only in number theory, but also in analysis and set theory. In fact, Hilbert writes in *The Logical Foundations of Mathematics*:

> Rather, our task is precisely to discover why and to what extent we always obtain correct results from the application of transfinite modes of inference of the sort that occur in analysis and set theory. The free use and the full mastery of the transfinite is to be achieved on the territory of finite! (Hilbert, 1923, p. 1140)

We then have the function of finitization. In fact, the formal theory represents, through the finite, the concepts and methods related to the infinite. Thus, in analysis and set theory, as in number theory, the mastery of the transfinite would be reached showing by finitistic methods the consistency of the corresponding formal systems that, as such, are of finite nature. Otherwise, symbolic knowledge would just be suppositive knowledge as Leibniz would say.

4   Concluding remarks

In this paper, a series of topics that are common to Leibniz and Hilbert have been emphasized. We did not intend to trace the history of influences (of the former over the latter) here, but to trace the history of ideas: the idea of mathematical knowledge obtained through the manipulation of signs present in both authors. There are obvious differences such as, for example, the respective notions of intuitive knowledge, but we recognize in Hilbert's account of mathematics some variants of the functions of the signs with which we describe Leibnizian symbolic knowledge. The by-product would be a better understanding of Hilbert's formalism.

References

Esquisabel, O. (2012), Representing and abstracting. An analysis of Leibniz's concept of symbolic knowledge, Vol. 41 of *Studies in Logic*, College Publications, pp. 1–49. In Lassale Casanave 2012b.

---

called "pragmatism of the variable": the objects that belong to the domain of variables should be given in advance in their identity and difference if we wish to ensure the reliable use of logical operations. (On Poincaré's pragmatism of the variable, see Heizmann, 1985). This restriction is overlooked in the algebraic conception of logical operations as a calculus.

Ewald, W. B. (1999), *From Kant to Hilbert*, Vol. II, Oxford University Press.

Ewald, W. B. & Sieg, W. (2013), *Introduction to David Hilbert's Lectures on the Foundations of Arithmetic and Logic 1917-1933*, Springer.

Frege, G. (1980), *Philosophical and Mathematical Correspondence*, Blackwell Publishers.

Heizmann, G. (1985), *Entre intuition et analyse. Poincaré et le concept de prédicativitité*, Librairie Scientifique et Technique Albert Blanchard.

Hibert, D. (1918), 'Axiomatisches Denken', *Mathematische Annalen* **78**, 405–415. English Translation by William Ewald in Ewald 1999, pp. 1105-1115.

Hilbert, D. (1923), 'Die logische Grundlagen der Mathematik', *Mathematische Annalen* **68**(1923), 151–165. English Translation by William Ewald in Ewald 1999 pp. 1134–48.

Hilbert, D. (1926), 'Über das Unendliche', *Mathematische Annalen* **95**(1926), 160–190. English Translation by Stefan Bauer Mengelberg in van Heijenoort 1967.

Hilbert, D. (1930), "Beweis des Tertium non Datur,' Nachrichten von der Gesellschaft der Wissenschaften zu Göttingen', *Mathematisch-Physikalische Klasse* **1930**, 120–125.

Hilbert, D. (1931), 'Die Grundlegung der elementaren Zahlenlehre', *Mathematische Annalen* **104**, 485–494. English Translation by William Ewald in Ewald 1999, pp. 1148-57.

Lassalle Casanave, A. (1995), Dos fundamentos à filosofia da aritmética Uma interpretação do Programa de Hilbert, PhD thesis, Unicamp.

Lassalle Casanave, A. (1996), 'Formalismo metodológico', *Papeles Uruguayos de Filosofía* **1**(1), 5–8.

Lassalle Casanave, A. (2010), Conocimiento simbólico, *in* J. C. Salles, ed., 'Empirismo e Gramática', Quarteto.

Lassalle Casanave, A. (2012*a*), Kantian avatars of symbolic knowledge. the role of symbolic manipulation in Kant's philosophy of mathematics, Vol. 41 of *Studies in Logic*, College Publications. In Lassale Casanave 2012b.

Lassalle Casanave, A., ed. (2012b), *Symbolic Knowledge from Leibniz to Husserl*, Vol. 41 of *Studies in Logic*, Col.

Leibniz, G. W. (1843-63), *Mathematische Schriften*, Vol. 1–7, Berlin und Halle. Ed. by C. I. Gerhardt, repr. by Georg Olms Verlag, Hildesheim/New York. 1971. Quoted as GM followed by volume and page number.

Leibniz, G. W. (1875-90), *Philosophische Schriften*, Vol. 1–7, Gerhardt. Ed. by C. I. Gerhardt, repr. by Georg Olms Verlag, Hildesheim/New York. 1978. Quoted as GP followed by volume and page number.

Leibniz, G. W. (1969), Philosophical papers and letters, D. Reidel Publishing Company. Ed. by Leroy E. Loemker. Quoted as Loemker 1969.

Shoenfield, J. R. (1967), *Mathematical Logic*, Addison-Wesley.

van Heijenoort, J. (1967), *From Frege to Gödel*, Harvard University Press.

Classical versus intuitionistic logic[‡]

Dag Prawitz*

* University of Stockholm
dag.prawitz@philosophy.su.se

1   Introduction

My interest in logic is first of all an interest in deductive reasoning. I see so-called classical first order predicate logic as an attempted codification of a fragment of inferences occurring in actual deductive practice. The ideal aim of such a codification is to create a deductive system — a regimented or formal language and rules of proofs — such that for all inferences accepted in actual practice there is a deduction in the system of a regimented sentence corresponding to the conclusion of the inference from regimented sentences corresponding to the premisses of the inference, and conversely, whenever there is a deduction in the system of a sentence from a set of premisses, informal practice would accept the corresponding deduction when translated back to natural language. As we know this ideal is too ambitious and we must be satisfied with a deductive system in which a fragment of actual inferences can be represented.

In the converse direction too one must expect discrepancies. Some inferences in the deductive system may have no correspondence in actual practice simply because they appear so trivial when translated to natural language, but this is of little concern as long as they would be accepted as correct if presented in practice. However, a codification of deductive practice — like most codifications of linguistic or legal practice — seems unavoidably to be more or less revisionary. Even a simple description of a practice has to disregard insignificant irregularities or mistakes in the practice. A codification has usually to go further and actually reform previous practice in order to achieve the desired systematization.

For instance, a system of natural deduction may take among its primitive inference rules the following ones

---

[‡]I thank professor Cesare Cozzo and professor Peter Schroeder-Heister for helpful comments to an earlier version of this essay.

$$\frac{A}{A \vee B} \qquad \frac{B}{A \vee B} \qquad \frac{A \vee B \quad \neg A}{B}$$

thereby making an arbitrary sentence B deducible from the two premises A and $\neg A$. The first two ones, disjunction introduction, are seldom made explicit, although they are certainly considered correct and may occur as an implicit step in other inferences, for instance, to infer C from A and $(A \vee B) \to C$. The third one is fairly common in informal practice and goes under several names, modus tollendo ponens or disjunctive syllogism. However, the inference of an arbitrary sentence B from the premises A and $\neg A$ is hardly sanctioned by informal practice — its correctness may even be doubted.

Trying to stay closer to actual practice in this respect, so-called relevance (or relevant) logicians have tried to develop codifications that do not contain this inference. Consequently, they have to depart from ordinary practice in some other respect, for instance by not allowing either disjunction introduction or the disjunctive syllogism or by somehow making deducibility intransitive. This illustrates how the systematization of our deductive practice seems to force us to modify the practice in one way or another.

The existence of several divergent codifications of informal deductive practice has naturally given rise to discussions of their relative virtues. I shall engage here in that kind of discussion with regard to classical and intuitionistic logic. Intuitionistic mathematics with its requirement that a proof of an existential statement should in principle establish the truth of one of its instances has of course an explicit revisionary aim. Intuitionistic first order predicate logic attempts to codify reasoning that satisfies this requirement, but strives otherwise to codify the same fragment of our deductive practice as first order classical logic.

I shall first attend to three well-known voices in the discussions of these two codifications and shall then turn to some methodological questions and to my own views on the issues involved.

Highly relevant to these discussions are the well-known translations of classical predicate logic into intuitionistic predicate logic, first discovered by Gentzen and Gödel. Also of some relevance is the (less well-known) translation of intuitionistic predicate logic into quantified classical S4 established by Prawitz & Malmnäs (1968). These translations will not be dealt with here. The emphasis will instead be on meaning-theoretical considerations, but they can be seen to some extent as spelling out the philosophical significance of the fact that classical logic can be translated into intuitionistic logic.

2   Some voices

2.1   Quine on orthodox and deviant logic

W. V. Quine, who devotes a chapter of his book *Philosophy of logic* to what he calls "deviant logics", voices a fairly common sentiment against intuitionistic logic compared with classical logic, saying that the former "lacks the familiarity, the convenience, the simplicity and the beauty of our logic". "Our logic" is according to Quine fixed by the definition of logical truth for compound sentences, which in turn is based on the classical truth (or satisfaction) conditions for sentences formed by the application of sentential operators and quantifiers. It is often said that these conditions determine the meanings of the sentential operators and the quantifiers, but Quine sees them like Tarski as clauses in a definition of truth and prefers not to use the notion of meaning.

Quine's main view of the deviant logician is that "he only changes the subject" — he is not opposing the laws of orthodox logic, but is talking about something else. For instance, if one stops to regard contradictions $A \& \neg A$ as implying all other sentences, then according to Quine (1970, pp. 81), one is not "talking about negation, '$\neg$', 'not'". Two different logics can therefore never be in conflict in the sense of making contrary assertions — they are simply talking about different things. In common (non-Quinean) parlance one may state this thesis by saying that two logics that come out with different sets of logical truths must attach different meanings to the logical constants.

2.2   Dummett on the philosophical basis of intuitionistic logic

Michael Dummett, who has no inhibition concerning the use of the term meaning, seemingly expresses the same view as Quine, saying "any disagreement over the validity of a logical law in which neither side is straightforwardly mistaken according to his own lights always reflects a divergence in the meanings each attaches to some or all the logical constants" (Dummett, 1991, pp. 193). But he makes an important qualification by adding immediately that this does not mean that such a disagreement is a mere verbal one. He develops his point by making a distinction between conceptually trivial and conceptually deep disagreements, the trivial ones being possible to resolve simply by introducing two different words and attaching different meanings to them, while the deep disagreements turn on different conceptions of what meaning is.

Only if it can be shown that the classical ideas about meaning are mistaken and that no intelligible meaning can be attached to the logical constants agreeing with the classical use of them, is the intuitionistic codification of mathematical reasoning of interest, according to Dummett. He accordingly dismisses an eclectic attitude that finds an interest in both the classical and the intuitionistic codification. It is true that a classical mathematician may also be interested in constructive proofs, but if one can grasp the sense that classical logic wants to attach to the logical constants, there is no particular reason, Dummett (1977) argues, to restrict oneself to the constructive methods recognized by intuitionistic logic; he refers to Markov's principle as an example of how one has access to constructive methods from a classical perspective that are not recognized by intuitionistic logic. To defend intuitionistic logic one must therefore not only argue for the positive thesis that intuitionistic logic is coherent but also for the negative thesis that classical logic is based on unintelligible and incoherent ideas.

As is well known, Dummett develops such arguments where he claims that knowledge of meaning must be publicly manifestable. Classical logic presupposes a bivalent notion of truth and the possibility of explaining the meaning of a logical constant c by the truth conditions of the sentences in which c occurs as the principal operator (the outermost symbol), but, he argues, some truth conditions of quantifications over infinite domains become knowledge transcendent, which rules out that knowledge of them can be manifestable. The conclusion is that the understanding of the logical constants imputed to us by classical logic is partly illusory.

2.3 Williamson on logical laws, meaning, and abductive methodology

Williamson (2007) constructs detailed example to show that two persons may disagree over a logical law and nevertheless mean the same with the logical constants occurring there, thus contradicting Dummett's and Quine's view that such disagreements imply differences with respect to what the persons are saying when using the logical constants. For instance, he constructs an example where two persons both use the English construction "if ... then ..." competently on the whole, thus supposedly meaning the same by the construction, but nevertheless disagree rationally about how the construction can be used in inferences, one supporting and the other rejecting certain instances of modus ponens.

Williamson agrees however with Quine in thinking that classical logic must be preferred to intuitionistic logic in view of such things as its greater elegance, simplicity, and strength. This judgement is part of an explicit abductive methodology (inference to the best explanation) on his part, where other virtues are explanatory power and consistency with known facts. Williamson concludes on the basis of such abductive arguments that classical logic is the right logic.[1] It should be said here that he does not see a logical system as primarily a codification of deductive reasoning in the way I am doing. For him logic is rather a discipline that tries to find absolute truths of a certain kind, overlapping with metaphysics, in principle not very unlike physics in resting on abductive methodology, although it does not use observations and experiments (Williamson, 2013).

3   Methodological considerations

How are issues of the kind considered above to be settled? It may be good to consider some aspects of this methodological question before arguing for a specific position on these issues.

3.1   First and third person perspective

A first question to raise concerns our own relation to the practice that we attempt to codify. We may adopt a first or a third person perspective. When one is not involved oneself in the practice that one attempts to codify, in other words, if the perspective is from a purely third person perspective, it is partly a descriptive or empirical enterprise. A codification that at least to some extent adopts a first person perspective is philosophically more interesting, because one is then faced with the question whether one wants to continue to make the inferences that one so far has been making, and one must ponder over the rationality of so doing. When a first person perspective is involved, I can usually rely on my own competence when considering the question whether an inference is accepted or not in actual practice. Furthermore, this empirical question becomes less urgent because it is eclipsed by the other question: am I really willing after reflection to draw the conclusion in question from the given premises; in other words, ought the inference belong to my continued practice.

This latter question gives rise to two further questions: What do I want to say with the sentences that occur in the inference, and what is it for the inference to be correct? The first one is partly conceptual or linguistic, and, if the notion of meaning can be made precise, it may be phrased as the question what I want my sentences to mean. Truth and warrant are notions that naturally come to mind when it comes to expound the second

---

[1] Hägerström's lectures at Uppsala University in 2014

question. Let us now turn to these questions of meaning and correctness of inferences in more detail.

3.2   Inference and meaning

As we have seen, Dummett, Quine, and Williamson all disagree on the notion of meaning and its impact on the validity of inferences. For Dummett the notion of meaning has a central position in the discussion of classical versus intuitionistic logic, while Williamson plays down its role in this connection, and Quine doubts that the notion is at all philosophical respectable; Quine's main point about unorthodox logic nevertheless depends somehow on meaning, and lacking this notion, he has to state his point in a somewhat diffused way, speaking about "the deviant logician's predicament: ...he only changes the subject".

It would be a mistake to take Williamson's examples to show linguistic meaning to be of little significance when discussing the validity of inference. They succeed at most in showing that what a person means by the sentences involved in an inference does not always *uniquely* determine whether she is accepting the inference as correct or not. One may grant this without denying that our way of understanding the logical constants plays a major part in our accepting an inference as valid and is frequently decisive for settling disagreements on such issues.

To take a simple example, a dispute about the validity of an inference

$$\frac{A \text{ or } B \qquad A}{\text{not } B}$$

which may be an example of Dummett's conceptually trivial disagreements, is likely settled by distinguishing between the exclusive and the inclusive meaning of "or". Almost every valid inference (the inference of A from A being an exception) would loose its validity by some change of what is meant by the involved words.

A first step when evaluating a codification of one's inferential practice is therefore reasonably to try to make up one's mind about what constants one wants to use and what they are to mean. The difficult and controversial question is how to make precise what a logical constant means.

3.3   Implicit and explicit knowledge

One question concerns what kind of knowledge one can have of meaning. According to Dummett, it must in the end be of an implicit kind. He argues that to avoid circularity, knowledge of meaning cannot be ascribed to a person ultimately, unless she somehow manifests the knowledge in her linguistic behaviour, typically in the case of the logical constants, by accepting or rejecting certain inferences as correct. Williamson's examples may be seen as directed in particular against Dummett's idea of how such implicit knowledge manifests itself.

This controversy is mainly concerned with knowledge of meaning from a third person perspective. However, when engaged in codifying one's own deductive practice, one must try to make explicit how the logical constants occurring in one's inferences are to be understood. It is true that this cannot be done without using some logical constants, and one problem is how to avoid getting into vicious circles by presupposing what is to be clarified.

Furthermore, it is to be noted that Williamson's main examples are concerned with knowledge of meaning ascribed to a person on the basis of her competence in speaking a natural language. To get anywhere when discussing different codifications of a deductive practice, we need however a much more precise notion of meaning for a regimented language.

3.4   Truth conditions

It is common to try to specify the meanings of sentences in terms of truth conditions. Quine objects to this as noted above. Following Tarski, he states truth conditions of compound sentences, not as a way to explain the logical constants, but as a first step in a definition of logical truth or logical consequence, which Quine takes to demarcate the logic that he is interested in. He points out that the truth conditions do not explain negation, conjunction, existential quantification and so on, because the conditions are using the corresponding logical constants and are thus presupposing an understanding of the very constants that they would explain. I think that he is essentially right in saying so and that the situation is even worse: when stating truth conditions, one is using an ambiguous natural language expression that is to be taken in a certain specific way, namely in exactly the sense that the truth condition is meant to specify.

Dummett points to a more basic reason why the truth conditions that occur in a definition of truth cannot function as meaning explanations: clearly they cannot simultaneously determine what truth is and the meaning of the involved logical constants. If truth is taken, not as defined by the truth conditions, but as an already understood notion, then the truth conditions make up what Dummett calls a modest theory of meaning, and his criticism of such a theory of meaning is essentially the same as Quine's. The circularity pointed out by Quine has the effect that both classical and intuitionistic logicians can agree about the truth-conditions, each one understanding the logical constants used in the meta-language in his or her own way. Consequently, the truth conditions are not able to illuminate how different logicians try to attach different meanings to the logical constants.

Truth conditions are important for Quine because, as mentioned, the ensuing definitions of logical truth and consequence constitute for Quine the way our logic is demarcated. It is a kind of codification of our inferences, although Quine does not use that word. However, as a demarcation or codification of the inferences that are to be taken as valid, it works as badly as a modest meaning theory. In order to decide whether an inference belongs to the codification, we have to reason in the meta-language, and how this reasoning goes will sometimes depend essentially on inferences whose inclusion in the codification is what we are trying to make decisions about. To codify reasoning by stating inference rules and axioms is thus superior, as everybody agrees, since in this way one reduces to a minimum the use of reasoning in the meta-language in order to see what the codification involves.

3.5 Meaning explained via inference rules

Given that truth conditions fail as meaning explanations and as a demarcation of our logic, one may ask if one should not see the stating of inference rules not as merely a codification of our deductive practice but also as a way to specify the meanings of the logical constants. A positive answer to this question, which seems first to have been proposed by Rudolf Carnap, is characteristic of what is now known as inferentialism.

Inferentialism can be of different kinds. What we may call radical inferentialism sees the meanings of sentences as constituted by the totality of inference rules that are in force. To specify the meaning of the regimented sentences cannot then, on this view, be a guide for how our inferential practice is to be codified properly, but becomes identical to such a codification.

The slogan "meaning is use", often ascribed to Ludwig Wittgenstein, has the same effect if taken to mean that the meaning of a sentence is determined by its total use. But what Wittgenstein says is rather that the meaning of a linguistic item can sometimes be explained by some of its use. This is a much more plausible thesis and seems to agree with ordinary experience: in some cases the meaning of an expression is adequately explained by describing its typical use, while in other cases we use the expression to make an assertion that is not trivially correct because of saying just what the expression means but that can be argued to follow from what the expression means and other facts.

The idea that the meaning of an expression is determined by some of its use was given a definite content by Dummett, who suggested that the meaning of an affirmative sentence is determined by its assertibility condition, that is, the condition that has to be satisfied if the assertion of the sentence is to be warranted.[2]

This suggestion can be seen as a generalization to the whole language of an idea that Gentzen (1935) had already formulated for the language of predicate logic in connection with his construction of a system of natural deduction. His proposal was that the meaning of a logical constant is determined by its introduction rules in that system. An introduction rule for a logical constant c sanctions inferences whose conclusions assert sentences A in which c is the principal operator. To say that certain rules are the introduction rules for the constant c is, however, not only to say that they are rules with conclusions of this form A, but is to say that proofs that terminate with an application of one of these rules are the direct or canonical proofs of sentences in question. By that is meant that all proofs of sentences where c is the principal operator can be reduced to proofs of this form, that is, proofs that terminate by an application of one the introduction rules.

The introduction rules for a logical constant c determine thereby the condition for asserting a sentence A in which c is the principal operator, in other words, what ground one must have in order to be right in asserting A. The required condition or ground is that one either has a canonical proof of A or knows how to reduce a certain non-canonical proof of A to canonical form. The idea is that the introduction rules give in this way the meanings of the sentences in question.

For this to be a non-circular explanation of the meaning of a sentence, the condition must not refer to sentences whose meanings are not yet explained. This requirement would not be satisfied, if we explained the meaning of set membership, that is, the meaning of sentences of the form $t \in \lambda x A(x)$, by giving the following introduction rule, corresponding to the principle of naïve set abstraction

$$\frac{A(t)}{t \in \lambda x A(x)}$$

The resulting condition for asserting the sentence $\lambda x(x \in x) \in \lambda x(x \in x)$ would be circular because in a canonical proof of this sentence the premiss of the last inference would be identical with the conclusion.

Gentzen's introduction rules satisfiy the requirement of non-circularity, because in the case of such a rule, a sentence that occurs as premiss or as

---

[2]This is also how Heyting (1956) explains the meaning of some logical constants.

discharged assumption is simply a sub-sentence of the sentence that occurs as conclusion. It is perfectly acceptable, however, that the introduction rule for a logical constant c is not pure[3] in the sense that it refers to another constant c^*, but then there must be an order of explanation between the constants in which the explanation of the meaning of c^* comes before that of c.

It is far from obvious that the meaning of an affirmative sentence can always be given in this way. Although one may certainly speak about the condition that has to be satisfied in order to be warranted to assert a sentence, the condition may not be possible to state in something like a non-circular rule. But when the meaning of a sentence A is explained in this way by a non-circular introduction rule, taken as the canonical way of proving A, it is made perfectly clear what is asserted by affirming A. In section 4, I shall consider to what extent classical and intuitionisticconstants can be explained in this way.

In contrast to what holds for radical inferentialism, to explain meaning in terms of introduction rules coincides only partially with codifying reasoning, since it states only the canonical ways of proving sentences. It can therefore obviously serve as a guide for which other inferences are to belong to the codification, a theme that we now turn to.

3.6 Correct inferences

When trying to codify one's own deductive practice, one will of course include in the codification only inferences that one thinks are correct. The correctness of an inference that does not discharge assumptions is naturally identified with it being the case that the conclusion follows from the premises. One may want to spell out "follows" here by saying that necessarily, if the sentences occurring in the premises are true, then so is the sentence occurring in the conclusion, which is how the notion of logical consequence is traditionally defined. This may lead one to agree with Williamson that logical questions are about truth and logical consequence and have to be settled by abductive reasoning.

However, even if one agrees with Williamson that people may rationally disagree about inferences although they agree about the meanings of the sentences involved, one cannot claim that the correctness of an inference is independent of the meaning attached to the logical constants. Unlike physics where one may think that one is studying given phenomena, logic is a field where the object of study, the correctness of inferences, is essentially created by our decisions about what the logical constants are

---

[3]A term introduced by Dummett (1991).

to mean. When we try to take a stand on the correctness of inferences, there are no given truths that we are to relate to in the same way as one may think that there are in physics, and therefore abductive reasoning seems inapplicable.

In model theory one applies instead deductive reasoning to given truth conditions to derive that a sentence is true or is a logical consequence of certain other sentences, but, as already remarked, the stating of truth conditions does not always differentiate between different meanings that we may want to attach to the logical constants, and to make logical consequence the criterion of correctness of inferences puts us in the awkward predicament that when pondering over an inference we may have to rely on reasoning that involves the inference whose correctness we are wondering about.

If we are able to fix the meanings of the logical constants by stating introduction rules in the way indicated above, we are in a better situation. It gives us a criterion when wondering about the correctness of an inference. If the inference is obtained by applying an introduction rule, it is of course trivially correct, because if we have proofs of the premises we have a canonical proof of the conclusion, which according to how the meaning of the sentence occurring in the conclusion is explained is sufficient for its assertion being warranted. If the inference is not an application of an introduction rule, then we ask whether proofs that have the inference as the last step can be reduced to canonical proofs. It is not that we never need to reason in order to decide whether this is the case. But as we shall see, at least classical and intuitionistic logicians will in crucial cases agree about the inferences used in this reasoning.

4 The meaning of the intuitionistic and the classical logical constants

Gentzen's introduction rules, taken as meaning constitutive of the logical constants of the language of predicate logic, agree, as is well known, with how intuitionistic mathematicians use the constants. On the one hand, the elimination rules stated by Gentzen become all justified when the constants are so understood because of there being reductions, originally introduced in the process of normalizing natural deductions, which applied to proofs terminating with an application of elimination rules give canonical proofs of the conclusion in question. On the other hand, no canonical proof of an arbitrarily chosen instance of the law of the excluded middle is known, nor any reduction that applied to a proof terminating with an application of the classical form of reductio ad absurdum gives a canonical proof of the conclusion. This gives a rationale for accepting Gentzen's intuitionistic system of natural deduction as a codification of reasoning where the logical constants are understood in the intuitionistic way, and supports what Dummett calls the positive intuitionistic thesis.

What is then to be said about the negative thesis that no coherent meaning can be attached on the classical use of the logical constants? Gentzen's introduction rules are of course accepted also in classical reasoning, but some of them cannot be seen as introduction rules, that is they cannot serve as explanations of meaning. The classical understanding of disjunction is not such that $A \vee B$ may be rightly asserted only if it is possible to prove either A or B, and hence Gentzen's introduction rule for disjunction does not determine the meaning of classical disjunction.

Similarly, an existential sentence $\exists x A(x)$ may be rightly asserted classically without knowing how to find a proof of some instance $A(t)$. Hence, Gentzen's introduction rule for the existential quantifier, which allows one to infer $\exists x A(x)$ from $A(t)$, does not determine what is to count classically as a canonical proof of $\exists x A(x)$ and therefore does not either determine the classical meaning of the existential quantifier.

This does not imply that the classical meanings of these constants cannot be explained in the same general way as the intuitionistic meanings of the logical constants have been explained. It is easy to see that appropriate introduction rules for the classical disjunctive connective and existential quantifier are given by the following schemata, where as usual assumptions of the form shown within square brackets are allowed to be discharged:

$$\frac{[\neg A, \neg B]}{\quad \bot \quad} \qquad \frac{[\forall x \neg A(x)]}{\quad \bot \quad}$$
$$\frac{\bot}{A \vee B} \qquad \frac{\bot}{\exists x A(x)}$$

These introduction rules are not pure but as was remarked above this is no hindrance since there is an order of explanation in which the other constants that occur in the schemata can be explained before the constants $\vee$ and $\exists$.

It is not possible to explain the classical use of negation in a similar way. To prove classically or intuitionistically the negation of a sentence A, we typically derive a contradiction from the assumption A. There is no other direct way of getting a ground for asserting $\neg A$ classically, although the classical use of negation differs from the intuitionistic. If we have in our language a special constant $\bot$ for absurdity or falsehood, the classical as well as the intuitionistic introduction rule for negation, $\neg I$, is as stated to the left below and justifies the elimination rule $\neg E$ stated to the right thereof, but in classical reasoning one also reasons according to the third schema for reductio (ad absurdum) shown to the right below:

$$\neg I) \quad \begin{array}{c} [A] \\ \bot \\ \hline \neg A \end{array} \qquad \neg E) \quad \frac{A \qquad \neg A}{\bot} \qquad \text{reductio)} \quad \begin{array}{c} [\neg A] \\ \bot \\ \hline A \end{array}$$

An inference of $\bot$ from the assumption $\neg A$ does not in general give a ground for asserting A when the meaning of negation is explained by taking its introduction rule to be the $\neg I$-rule. It is possible however to validate this form of reduction in classical reasoning by taking atomic sentences to be understood classically in a particular way. Given an intuitionistic one-place predicate P_i, we can introduce a classical predicate P_c and explain its meaning by letting its introduction rule be

$$\begin{array}{c} [\neg P_i(t)] \\ \bot \\ \hline P_c(t) \end{array}$$

We get a codification of classical reasoning based on meaning explanations of the same kind as we got for intuitionistic reasoning, by adding classical predicates with introduction rules in the way just exemplified, by adopting the above introduction rules for $\vee$ and $\exists$, and by taking the introduction rules for the other logical constants to be the ones stated by Gentzen. Classical reasoning is in this way shown to cohere fully with meaning explanations just as intuitionistic reasoning, which contradicts Dummett's negative thesis.

This is of course not how the meanings of the classical constants are usually explained or understood. One could say that the classical understanding of disjunction is such that Gentzen's simple introduction rules for disjunction are immediately justified and do not need to be derived from the more complicated classical introduction rule stated above. But this is a minor point, and one could declare that classical disjunction have three introduction rules, Gentzen's two rules and the rule stated here — it is mostly a question of elegance not to adopt Gentzen's rules as primitive, since they hold as derived rules when the above $\vee I$-rule is adopted.

To explain the classical meanings of predicates via intuitionistic ones is certainly very foreign to usual classical conceptions, but the question we are dealing with here is whether, after having rejected classical meaning explanations in terms of truth conditions, it is possible to understand classical reasoning as based coherently on some other explanations of meaning. The explanations suggested here show this to be possible.

5 A codification with both classical and intuitionistic constants mixed together

Having seen that the classical as well as the intuitionistic codification of deductive practice is fully justified on the basis of different meanings attached to the involved expressions, a choice between the codifications should reasonably depend on what one wants to say with one's sentences. For instance, if one is interested in the computational content of an inferred existential sentence $\exists x A(x)$ and wants it to say that an assertible instance $A(t)$ can be found, then one should choose the intuitionistic codification.[4] If one does not care about this and is satisfied with the weaker existential assertion provided by the classical codification, one can still choose the intuitionistic codification since this assertion is also available there by using the sentence $\neg\forall x \neg A(x)$, but one may as well choose the classical codification as more convenient.

Comparing the two codifications, it is clearly wrong to argue that classical logic is stronger than intuitionistic. What can be said is instead that the intuitionistic language is more expressive than the classical one, having access to stronger existence statements that cannot be expressed in the classical language. However, there is no need to choose between the two codifications because we can have a more comprehensive one that codifies both classical and intuitionistic reasoning based on a uniform pattern of meaning explanations.

When the classical and intuitionistic codifications attach different meanings to a constant, we need to use different symbols, and I shall use a subscript c for the classical meaning and i for the intuitionistic. The classical and intuitionistic constants can then have a peaceful coexistence in a language that contains both.

Let us consider a language with the constants $\bot, \neg, \vee$, and $\forall$ that are common for classical and intuitionistic logic, the particular classical logical constants $\vee_c, \to_c$, and $\exists_c$, the particular intuitionistic logical constants $\vee_i, \to_i$, and $\exists_i$, and predicates with subscripts i or c.

Letting the codification take the form of a natural deduction system, it is to contain

1. Gentzen's introduction and elimination rules for $\bot, \neg, \wedge$ and $\forall$ (the introduction rule for $\bot$ is vacant and the elimination rule allows the inference of an arbitrary sentence from $\bot$);

---

[4]As was pointed out by Dummett, if one has this interest, one may ask if the intuitionistic codification should not be enriched with Markov's principle. It is an interesting question that I have not been able to attend to here.

2. Gentzen's introduction and elimination rules for $\lor, \to$, and $\exists$ where now i is attached as a subscript to the constant;

3. the following introduction and elimination rule for $\lor_c, \to_c,$ and $\exists_c$

$$
\frac{\begin{array}{c}[\neg A, \neg B]\\ \bot\end{array}}{A \lor_c B}
\qquad
\frac{\begin{array}{c}[A, \neg B]\\ \bot\end{array}}{A \to_c B}
\qquad
\frac{\begin{array}{c}[\forall x \neg A(x)]\\ \bot\end{array}}{\exists_c x A(x)}
$$

$$
\frac{A \lor_c B \qquad \neg A \qquad \neg B}{\bot}
\qquad
\frac{A \to_c B \qquad A \qquad \neg B}{\bot}
$$

$$
\frac{\exists_c x A(x) \qquad \forall x \neg A(x)}{\bot}
$$

4. introduction rules for predicates P_c of the kind already exemplified above, and the corresponding elimination rules exemplified by

$$
\frac{P_c(t) \qquad \neg P_i(t)}{\bot}
$$

In addition there may be particular introduction and elimination rules for a predicate P_i. Sometimes, for instance when P_i is the predicate N of being a natural number, the classical rules for P_c follow as derived rules and then we do not need to differentiate between the classical and the intuitionistic predicate.

A sentence that contains only classical constants (including the ones common for classical and intuitionistic logic) comes out as provable in this mixed system if and only if it is provable in classical logic, and a sentence that contains only intuitionistic constants comes out as provable in the mixed system if and only if it is provable in intuitionistic logic.

The relative strength of the classical and intuitionistic constants become visible in the mixed system, where $A \lor_c B, A \to_c B, \exists_c x A(x)$ and $P_c(t)$ can be deduced from $A \lor_i B, A \to_i B, \exists_i x A(x)$ and $P_i(t)$, respectively, but not vice versa.

The view voiced by Quine that the different codifications shall not be seen as being in conflict with each other is supported here, but in a way quite different from how he was thinking. The classical logician is not asserting what the intuitionistic logician denies. For instance, the classical logician asserts $A \lor_c \neg A$ to which the intuitionist does not object; he objects to the universal validity of $A \lor_i \neg A$, which is not asserted by the classical logician.

If they are sufficiently ecumenical and can use the other's vocabulary in their own speech, a classical logician and an intuitionist can both adopt the present mixed system, and the intuitionist must then agree that $A \vee_c \neg A$ is trivially provable for any sentence A, even when it contains intuitionistic constants, and the classical logician must admit that he has no ground for universally asserting $A \vee_i \neg A$, even when A contains only classical constants. That would require a general method for finding for any A a canonical proof of $A \vee_i \neg A$ whose immediate sub-proof must be either a proof of A or a proof of $\neg A$, and we do not know any such method.

In the case of the classical form of reductio, the situation is somewhat different. The only constants explicitly involved here are negation and falsehood, understood in the same way classically and intuitionistically, so the classical and intuitionistic logicians are now speaking about the same thing. The intuitionist must agree to such an inference when the inferred sentence contains only classical constants. The classical logician, who initially endorses the inference schema as universally correct, must retract when he realizes that the inferred sentence A may contain one of the particular intuitionistic constants that is not common with the classical ones. Although he is still speaking about the common negation, he now agrees that when A contains an intuitionistic constant, he cannot always infer A after having derived $\bot$ from $\neg A$.

It has sometimes been held that a deductive system that contains both classical and intuitionistic constants is impossible, because the different constants would collapse. Popper (1948) was the first to observe that in a system containing the following rules for classical and intuitionistic negation (now formulated without $\bot$)

$$\dfrac{\begin{array}{cc}[A] & [A] \\ B & \neg B\end{array}}{\neg_i A} \qquad \dfrac{A \qquad \neg_i A}{B} \qquad \dfrac{\begin{array}{cc}[A] & [A] \\ B & \neg B\end{array}}{\neg_c A} \qquad \dfrac{\begin{array}{cc}[\neg_c A] & [\neg_c A] \\ B & \neg_c B\end{array}}{A}$$

where A and B may be arbitrary sentences, the two negations collapse into one, that is, $\neg_i A$ and $\neg_c A$ become derivable from each other.[5] He had an idea roughly described by saying that a logical connective is characterized by the inference rules that hold for it. Thus, the meaning of intuitionistic negation is given by the first two rules above, and the meaning of classical negation by the last two rules above, and one may think that one has thereby characterized two different negations adequately in accordance with classical and intuitionistic reasoning. But since they collapse in a system that contains both negations, we get the result that the classical reductio holds also intuitionistically, flatly contradicting intuitionistic

---

[5] See also Schröeder-Heister (1984).

reasoning. Popper's idea is a form of what I called radical inferentialism above, which I think must be rejected for the reasons that were hinted to.

In contrast, from the point of view advocated here, it is immediately clear that classical and intuitionistic negation coincide since they have the same I-rule and that this rule justifies the second but not the fourth rule above when A and B are arbitrary sentences.

The collapse noted by Popper is a special case of the general and easily verified fact that two constants that both satisfy a Gentzen pair of I- and E-rules (introduction and elimination rules) are deductively equivalent — instances of Gentzen's $\neg E$-rule or of the second rule above are also instance of the classical reductio or of the fourth rule above, respectively. This fact may lead one to think that the special classical and intuitionistic constants would collapse. They all satisfy Gentzen's I- and E-rules when the codifications are kept as separate systems, and one may think that they should do so also in the codification that mixes the constants. But even in the separate classical system it would be wrong to say, for instance, that Gentzen's $\vee I$-rule determines the meaning of classical disjunction, and as becomes clear in the codification that comprises both classical and intuitionistic disjunction, $\vee_c$ is weaker than $\vee_i$ when it is given its proper I-rule. Although Gentzen's $\vee I$-rule therefore holds also for $\vee_c$, his $\vee E$-rule is not generally justified for $\vee_c$ when the meaning of $\vee_c$ is taken to be determined by the $\vee_c I$-rule. However, the rule does hold if its minor premiss C is stable, that is, if C and $\neg\neg C$ are deductively equivalent.

References

Dummett, M. (1977), *Elements of Intuitionism*, Clarendon Press.

Dummett, M. (1991), *The Logical Basis of Metaphysics*, Duckworth.

Gentzen, G. (1935), 'Untersuchungen über das logische schliessen', *Mathematische Zeitschrift* **39**, 176–210.

Popper, K. (1948), 'On the theory of deduction, part ii: The definitions of classical and intuitionist negation', *Indagationes mathematicæ* **10**, 111–120.

Prawitz, D. & Malmnäs, P.-E. (1968), *Contributions to Mathematical Logic*, North-Holland, chapter A survey of some connections between classical, intuitionistic and minimal logic, pp. 215–229.

Quine, W. V. (1970), *The Philosophy of Logic*, Prentice-hall.

Schröeder-Heister, P. (1984), 'Popper's theory of deductive inference and the concept of logical concept', *History and Philosophy of Logic* 5, 79–110.

Williamson, T. (2007), *The Philosophy of Philosophy*, Blackwell.

Williamson, T. (2013), *Modal Logic as Metaphysics*, Oxford University Press.

Experiments with negation

Frank Thomas Sautter*

* Departamento de Filosofia
Universidade Federal de Santa Maria
Conselho Nacional de Desenvolvimento Científico e Tecnológico
ftsautter@ufsm.br

1   Introduction

In this work in honor of Luiz Carlos Pinheiro Dias Pereira I conduct two experiments with negation. Both experiments are philosophically motivated, and both are related to research and to works supervised by Luiz Carlos.

In the first experiment I obtained relations of logical opposition in a First-Order Predicate Logic in which negation applied to open formulas does not satisfy *Tertium Non Datur*, while negation applied to closed formulas (sentences) satisfies *Tertium Non Datur*. This simulates, in a single and contemporary framework, the traditional syllogistic discussion on the relations between judicative and predicative negation. Concerning categorical syllogistic, Pereira et al. (2008) demonstrated, as a corollary of results in constructive fragments of Classical First-Order Predicate Logic, its constructive character, although the process of reduction of valid modes is not constructive. Regarding predicative negation and related topics, da Fonseca (2007) wrote a Master's Dissertation on infinite judgements in Kant's Critique of Pure Reason, under the direction of Luiz Carlos.

Translations between logics is a relatively new speciality of the Brazilian School of Logic. It is a valuable research area, mainly due to its practical consequences, e.g., the transfering of metatheoretical results from a logic to another. In the second experiment I designed three logical systems inspired by well-known translations of Classical Propositional Logic into Intuitionistic Propositional Logic: Kolmogorov Logic results of a mapping between formulas in which double negation is applied to all subformulas, proper or improper, of a formula; Glivenko Logic results of a mapping between formulas in which double negation is applied only to the improper subformula of a formula; and Gentzen Logic results of a mapping between formulas in which double negation is applied only to the atomic subformulas of a formula. I tried to provide finite valued semantics to all of them. To prevent an atomic sentence to be derivable from itself (which is not allowed in any of the three logics), these semantics show an unusual feature, namely, there are more values attributable to the molecular sentences than to the atomic ones[1]. Unfortunately this type of solution does not work, due to the phenomena of vacuity. An appropriate solution must rely on global conditions. Paiva & Pereira (2006) present an example of work by Luiz Carlos with respect to Intuitionistic Logic.

2   An experiment inspired by Syllogistic

As is well known, obversion is an immediate inference in which from a given categorical proposition another categorical proposition is inferred whose subject is the same, whose predicate is the original one negated, and whose quality is changed. So, the obverted of "All S are P" is "No S are $nonP$", the obverted of "No S are P" is "All S are $nonP$", the obverted of "Some S are P" is "Some S are not $nonP$", and the obverted of "Some S are not P" is "Some S are $nonP$".

Aristotle expressly rejects at least some forms of obversion. In the Book 1, Chapter XLVI of the *Prior Analytics* he says:

> It makes no little difference in establishing or refuting a proposition whether we suppose that 'not to be so-and-so' and 'to be not-so-and-so' mean the same or something different: e.g., whether 'not to be white' means the same as 'to be not-white'. For it does not mean the same; the negation of 'to be white' is not 'to be non-white' but 'not to be white' (Aristotle, 1938, p. 395).

Let us make the following distinction between an use of negation applied to open formulas and an use of negation applied to closed formulas (sentences). Negation applied to sentences behaves like classical negation: for every sentence φ, $\varphi \vee \neg\varphi$ is valid. But negation applied to open formulas behaves like a constructive negation. If an open formula φ has n free variables, the n-tuples of individuals from the domain of discourse are divided into three classes, namely, the class of n-tuples that satisfies φ,

---

[1]In a diametrically opposite direction, Bochvar (1981) designed an external connectives system in which values distinct from the True and the False are obliterated by the application of the connectives.

the class of n-tuples that does not satisfy φ, and the class of n-tuples that neither satisfies nor does not satisfy φ. Negation applied to open formulas does not satisfy *Tertium Non Datur*: not for every open formula φ with n free variables $x_1, \ldots, x_n, \forall x_1, \ldots, x_n(\varphi(x_1, \ldots, x_n) \vee \neg\varphi(x_1, \ldots, x_n))$.

Remark 1. *Let φ be an open formula with a single free variable x. If $\forall x\varphi(x)$ is true, $\neg\exists x\neg\varphi(x)$ is also true.*

Proof. If $\forall x\varphi(x)$ is true, the class of individuals that satisfies φ equals the domain of discourse, the class of individuals that does not satisfy φ is empty, $\exists x\neg\varphi(x)$ is false, and $\neg\exists x\neg\varphi(x)$ is true. $\qquad\square$

As a corollary it follows that:

Remark 2. *Let φ be an open formula with a single free variable x. If $\forall x\varphi(x)$ is true, $\exists x\neg\varphi(x)$ is false.*

Remark 3. *There is an open formula φ with a single free variable x such that $\neg\exists x\neg\varphi(x)$ is true but $\forall x\varphi(x)$ is false.*

Proof. If the class of individuals that does not satisfy φ is empty and the class of individuals that neither satisfies nor does not satisfy φ is not empty, $\neg\exists x\neg\varphi(x)$ is true but $\forall x\varphi(x)$ is false. $\qquad\square$

As a corollary it follows that:

Remark 4. *There is an open formula φ with a single free variable x such that $\exists x\neg\varphi(x)$ is false and $\forall x\varphi(x)$ is also false.*

Remark 5. *Let φ be an open formula with a single free variable x. If $\exists x\neg\varphi(x)$ is true, $\neg\forall x\varphi(x)$ is also true.*

Proof. If $\exists x\neg\varphi(x)$ is true, the class of individuals that does not satisfy φ is not empty, the class of individuals that satisfies φ does not equal the domain of discourse, $\forall x\varphi(x)$ is false, and $\neg\forall x\varphi(x)$ is true. $\qquad\square$

As a corollary it follows that:

Remark 6. *Let φ be an open formula with a single free variable x. If $\neg\exists x\neg\varphi(x)$ is false, $\neg\forall x\varphi(x)$ is true.*

Remark 7. *There is an open formula φ with a single free variable x such that $\neg\forall x\varphi(x)$ is true but $\exists x\neg\varphi(x)$ is false.*

Proof. If the class of individuals that does not satisfy φ is empty and the class of individuals that neither satisfies nor does not satisfy φ is not empty, $\neg\forall x\varphi(x)$ is true but $\exists x\neg\varphi(x)$ is false. $\qquad\square$

As a corollary it follows that:

Remark 8. *There is an open formula φ with a single variable x such that $\neg\forall x\varphi(x)$ is true and $\neg\exists x\neg\varphi(x)$ is also true.*

Remark 9. *Let φ be an open formula with a single free variable x. If $\exists x\varphi x$ is true, $\neg\forall x\neg\varphi x$ is also true.*

Proof. If $\exists x\varphi(x)$ is true, the class of individuals that satisfies φ is not empty, the class of individuals that does not satisfy φ does not equal the domain of discourse, $\forall x\neg\varphi(x)$ is false, and $\neg\forall x\neg\varphi(x)$ is true. □

As a corollary it follows that:

Remark 10. *Let φ be an open formula with a single free variable x. If $\exists x\varphi(x)$ is true, $\forall x\neg\varphi(x)$ is false.*

Remark 11. *There is an open formula φ with a single free variable x such that $\neg\forall x\neg\varphi(x)$ is true and $\exists x\varphi(x)$ is also true.*

Proof. If the class of individuals that satisfies φ is not empty, $\neg\forall x\neg\varphi(x)$ and $\exists x\varphi(x)$ are truth. □

As a corollary it follows that:

Remark 12. *There is an open formula φ with a single free variable x such that $\forall x\neg\varphi(x)$ is false but $\exists x\varphi(x)$ is true.*

Remark 13. *Let φ be an open formula with a single free variable x. If $\forall x\neg\varphi(x)$ is true, $\neg\exists x\varphi(x)$ is also true.*

Proof. If $\forall\neg\varphi(x)$ is true, the class of individuals that does not satisfy φ equals the domain of discourse, the class of individuals that satisfies φ is empty, $\exists\varphi(x)$ is false, and $\neg\exists x\varphi(x)$ is true. □

As a corollary it follows that:

Remark 14. *Let φ be an open formula with a single free variable x. If $\neg\forall x\neg\varphi(x)$ is false, $\neg\exists x\varphi(x)$ is true.*

Remark 15. *There is an open formula φ with a single free variable x such that $\neg\exists x\varphi(x)$ is true but $\forall x\neg\varphi(x)$ is false.*

Proof. If the class of individuals that satisfies y φ is empty and the class of individuals that neither satisfies nor does not satisfy φ is not empty, $\neg\exists x\varphi(x)$ is true but $\forall x\neg\varphi(x)$ is false. □

As a corollary it follows that:

Remark 16. *There is an open formula φ with a single free variable x such that $\neg\exists x\varphi(x)$ is true and $\neg\forall x\neg\varphi(x)$ is also true.*

Figure 1 shows a square of oppositions with homogeneous classical negation, i.e., with classical negation applied to sentences as well as to open formulas. At each corner there is a pair of logically equivalent sentences. An arrow from α to β indicates that a certain truth value of α forces a certain truth value of β.

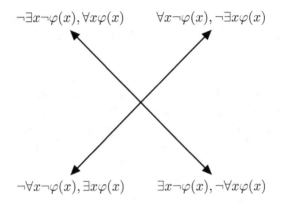

$$\neg\exists x\neg\varphi(x), \forall x\varphi(x) \qquad \forall x\neg\varphi(x), \neg\exists x\varphi(x)$$

$$\neg\forall x\neg\varphi(x), \exists x\varphi(x) \qquad \exists x\neg\varphi(x), \neg\forall x\varphi(x)$$

Figure 1: Classical square of contradictories

Figure 2 shows the expansion of the diagonal relation of Figure 1 when sentence negation is classical and open formula negation is constructive. The diagonal relation as well as the antidiagonal relation are relations of contradiction, because sentence negation is classical. Remarks 1 and 3 establish that the left edge is a relation of subalternation. Remarks 5 and 7 establish that the right edge is also a relation of subalternation. Remarks 2 and 4 establish that the superior edge is partially a relation of contradiction; the prefix 'sub' indicates a change of quality, from an universal to a particular. Remarks 6 and 8 establish that the inferior edge is also partially a relation of contradicition.

Figure 3 shows the expansion of the antidiagonal relation of the Figure 1 when sentence negation is classical and open formula negation is constructive. The diagonal relation [2] as well as the antidiagonal relation [3]

---

[2] The relation between the formula from the top left corner of the square and the formula from the bottom right corner of the square.

[3] The relation between the formula from the the top right corner of the square and the formula from the bottom left corner of the square.

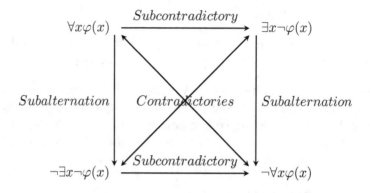

Figure 2: Square of the diagonal contradictories

are relations of contradiction, because sentence negation is classical. Remarks 9 and 11 establish that the left edge is a relation of subalternation. Remarks 13 and 15 establish that the right edge is also a relation of subalternation. Remarks 10 and 12 establish that the superior edge is partially [4] a relation of contradiction; the prefix 'super' indicates a change of quality, from a particular to an universal. Remarks 14 and 16 establish that the inferior edge is also partially a relation of contradiction.

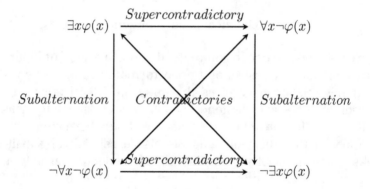

Figure 3: Square of the antidiagonal contradictories

---

[4]Only half of the truth conditions of the relation of contradiction is present.

3 An experiment inspired by translations between logics

Kolmogorov (1967), Glivenko (1929), and Gentzen (1936) proposed translations from Classical Propositional Logic (CPL) into Intuitionistic Propositional Logic. First of all, let us see what such translations are.

Let $\mathcal{L}(\neg, \supset)$ be the language of CPL restricted to negation and implication, and $\mathcal{F}(\mathcal{L}(\neg, \supset))$ be the set of formulas of $\mathcal{L}(\neg, \supset)$.

Kolmogorov (1967) gives the following translation $*$ from $\mathcal{F}(\mathcal{L}(\neg, \supset))$ into itself:

- $\varphi^* = \neg\neg\varphi$, for φ atomic,

- $(\neg\varphi)^* = \neg\neg\neg\varphi^*$,

- $(\varphi \supset \psi)^* = \neg\neg(\varphi^* \supset \psi^*)$.

He proves the following result:

If $\Gamma \vdash_{\mathcal{H}} \varphi$ then $\Gamma^* \vdash_{\mathcal{B}} \varphi^*$, where $\Gamma^* = \{\psi^* : \psi \in \Gamma\}$; $\mathcal{H}$, for 'Hilbert', is CPL; and $\mathcal{B}$, for 'Brouwer', is an Intuitionistic Propositional Logic.

The Kolmogorov translation applies double negation to all the subformulas of a formula.

Glivenko (1929) proves the following results:

1. If φ is provable in CPL, then $\neg\neg\varphi$ is provable in Intuitionistic Propositional Logic.

2. If $\neg\varphi$ is provable in CPL, then it is also provable in Intuitionistic Propositional Logic.

Glivenkos results suggests the following translation $**$ from $\mathcal{F}(\mathcal{L}(\neg, \supset))$ into itself: $\varphi^{**} = \neg\neg\varphi$. This translation applies double negation only to the improper subformulas of a formula, not to its proper subformulas.

If we restrict the language to $\mathcal{L}(\neg, \supset)$, Gentzens translation (Gentzen, 1936) applies double negation only to the atomic subformulas of a formula.

Let us try some experiments with these three translations.

The Glivenko mapping g: $\mathcal{F}(\mathcal{L}(\neg, \supset)) \to \mathcal{F}(\mathcal{L}(\neg, \supset))$ is given by the following condition: $g(\varphi) = \neg\neg\varphi$.

The consequence relation of the Glivenko Logic (GL), $\vdash_{GL}$, is given by the following condition: $\Gamma \vdash_{GL} \varphi$ iff $\Theta \vdash_{CPL} \psi$, where $\Gamma = \{g(\gamma) : \gamma \in \Theta\}$ and φ is $g(\psi)$.

The operator $\mathcal{C}_{GL}$ associated with $\vdash_{GL}$ (Γ), i.e., $\varphi \in \mathcal{C}_{GL}$ iff $\Gamma \vdash_{GL} \varphi$, restricted to $g(\mathcal{F}(\mathcal{L}(\neg, \supset)))$, is a Tarskian consequence operator.

The Kolmogorov mapping k: $\mathcal{F}(\mathcal{L}(\neg, \supset)) \to \mathcal{F}(\mathcal{L}(\neg, \supset))$ is given by the following conditions:

- $k(\varphi) = \neg\neg\varphi$, for φ atomic,

- $k(\neg\varphi) = \neg\neg\neg\, k(\varphi)$, for φ atomic or if $\supset$ is the main connective of φ,

- $k(\neg\varphi) = \neg\, k(\varphi)$, if $\neg$ is the main connective of φ.

- $k(\varphi \supset \psi) = \neg\neg\, (k(\varphi) \supset k(\psi))$.

The consequence relation of the Kolmogorov Logic (KL), $\vdash_{KL}$, is given by the following condition: $\Gamma \vdash_{KL} \varphi$ iff $\Theta \vdash_{CPL} \psi$, where $\Gamma = \{k(\gamma) : \gamma \in \Theta\}$ and φ is $k(\psi)$.

The operator $\mathcal{C}_{KL}$ associated with $\vdash_{KL}$ (Γ), i.e., $\varphi \in \mathcal{C}_{KL}$ iff $\Gamma \vdash_{KL} \varphi$, restricted to $k(\mathcal{F}(\mathcal{L}(\neg, \supset)))$, is a Tarskian consequence operator.

The Gentzen mapping t: $\mathcal{F}(\mathcal{L}(\neg, \supset)) \to \mathcal{F}(\mathcal{L}(\neg, \supset))$ is given by the following conditions:

- $t(\varphi) = \neg\neg\varphi$, for φ atomic,

- $t(\neg\varphi) = \neg\, t(\varphi)$,

- $t(\varphi \supset \psi) = t(\varphi) \supset t(\psi)$.

The consequence relation of the Gentzen Logic (TL), $\vdash_{TL}$, is given by the following condition: $\Gamma \vdash_{TL} \varphi$ iff $\Theta \vdash_{CPL} \psi$, where $\Gamma = \{t(\gamma) : \gamma \in \Theta\}$ and φ is $t(\psi)$.

The operator $\mathcal{C}_{TL}$ associated with $\vdash_{TL}$ (Γ), i.e., $\varphi \in \mathcal{C}_{TL}$ iff $\Gamma \vdash_{TL} \varphi$, restricted to $t(\mathcal{F}(\mathcal{L}(\neg, \supset)))$, is a Tarskian consequence operator.

None of these three logics can have a finite valued semantics in which a designated value is assigned to atomic formulas. If they did, they would not have a correct semantics because $\{\varphi\} \nvdash_{GL} \varphi$ ($\{\varphi\} \nvdash_{KL} \varphi$, and $\{\varphi\} \nvdash_{TL} \varphi$), for φ atomic.

A suggestive design of finite valued semantics for these three logics comprises dividing values into levels: Level A consists of the values 0 and 1, Level B consists of the values 2 and 3, Level C consists of the values 4 and 5, and Level D consists solely of the value 6. Double negation allows the rise from Level A to Level C, passing through Level B. Level D corresponds to the level of rejected formulas, i.e., formulas which do not enter the relation of consequence. At each level, except Level D, an even value corresponds to the False and an odd value correspond to the True. The value 5, i.e., the True at the Level C is the only designated value.

The following finite valued semantics is a proposal for Glivenko Logic:

Valuations for negation:

- $v(\neg\varphi) = 2$, if $v(\varphi) = 1$,

- $v(\neg\varphi) = 3$, if $v(\varphi) = 0$,

- $v(\neg\varphi) = 4$, if $v(\varphi) = 3$ or $v(\varphi) = 5$,

- $v(\neg\varphi) = 5$, if $v(\varphi) = 2$ or $v(\varphi) = 4$.

Valuations for implication:

- $v(\varphi \supset \psi) = 0$, if $v(\varphi)$ is odd and $v(\psi)$ is even,

- $v(\varphi \supset \psi) = 1$, if $v(\varphi)$ is even or $v(\psi)$ is odd.

Similarly, the following finite valued semantics is a proposal for Kolmogorov Logic:

Valuations for negation:

- $v(\neg\varphi) = 2$, if $v(\varphi) = 1$,

- $v(\neg\varphi) = 3$, if $v(\varphi) = 0$,

- $v(\neg\varphi) = 4$, if $v(\varphi) = 3$ or $v(\varphi) = 5$,

- $v(\neg\varphi) = 5$, if $v(\varphi) = 2$ or $v(\varphi) = 4$,

- $v(\neg\varphi) = 6$, if $v(\varphi) = 6$.

Valuations for implication:

- $v(\varphi \supset \psi) = 0$, if $v(\varphi) = 5$ and $v(\psi) = 4$,

- $v(\varphi \supset \psi) = 1$, if $v(\varphi) \in \{4,5\}$ and $v(\psi) = 5$, or $v(\varphi) = v(\psi) = 4$,

- $v(\varphi \supset \psi) = 6$, if $v(\varphi) \notin \{4,5\}$ or $v(\psi) \notin \{4,5\}$.

Likewise, the following finite valued semantics is a proposal for Gentzen Logic:

Valuations for negation:

- $v(\neg\varphi) = 2$, if $v(\varphi) = 1$,

- $v(\neg\varphi) = 3$, if $v(\varphi) = 0$,

- $v(\neg\varphi) = 4$, if $v(\varphi) = 3$ or $v(\varphi) = 5$,

- $v(\neg\varphi) = 5$, if $v(\varphi) = 2$ or $v(\varphi) = 4$,

- $v(\neg\varphi) = 6$, if $v(\varphi) = 6$.

Valuations for implication:

- $v(\varphi \supset \psi) = 4$, if $v(\varphi) = 5$ and $v(\psi) = 4$,

- $v(\varphi \supset \psi) = 5$, if $v(\varphi) \in \{4,5\}$ and $v(\psi) = 5$, or $v(\varphi) = v(\psi) = 4$,

- $v(\varphi \supset \psi) = 6$, if $v(\varphi) \notin \{4,5\}$ or $v(\psi) \notin \{4,5\}$.

The three finite valued semantics above appear to be good solutions, but they all suffer from the same kind of problem: a kind of vacuity. In CPL we derive any formula from an inconsistent set of formulas, formulas which cannot be satisfied, formulas for which there cannot simultaneously be assigned the designated value. In particular, one cannot assign the designated value to logical falsehoods. Atomic formulas of these three logics behave like a logical falsehood, because one cannot assign the designated value to them.

There are two solutions to this problem.

Perhaps the easiest way to deal with this kind of problem is to limit the language of CPL. If $\mathcal{L}(\neg, \supset)$ is the language of CPL, then:

1. The restricted language $\mathcal{L}_{GL}(\neg, \supset)$ of Glivenko Logic is given by the following condition: if $\varphi \in \mathcal{L}(\neg, \supset)$, then $\neg\neg\varphi \in \mathcal{L}_{GL}(\neg, \supset)$.

2. The restricted language $\mathcal{L}_{KL}(\neg, \supset)$ of Kolmogorov Logic is given by the following conditions:

    (a) If $\varphi \in \mathcal{L}(\neg, \supset)$ is atomic, $\neg\neg\varphi \in \mathcal{L}_{KL}(\neg, \supset)$ and $\neg\neg\neg\varphi \in \mathcal{L}_{KL}(\neg, \supset$).

    (b) If $\psi \in \mathcal{L}_{KL}(\neg, \supset)$ and $\theta \in \mathcal{L}_{KL}(\neg, \supset)$, then $\neg\neg\neg\psi \in \mathcal{L}_{KL}(\neg, \supset)$ and $\neg\neg(\psi \supset \theta) \in \mathcal{L}_{KL}(\neg, \supset)$.

3. The restricted language $\mathcal{L}_{TL}(\neg, \supset)$ of Gentzen Logic is given by the following conditions:

    (a) If $\varphi \in \mathcal{L}(\neg, \supset)$ is atomic, $\neg\neg\varphi \in \mathcal{L}_{TL}(\neg, \supset)$.

    (b) If $\psi \in \mathcal{L}_{TL}(\neg, \supset)$ and $\theta \in \mathcal{L}_{TL}(\neg, \supset)$, then $\neg\psi \in \mathcal{L}_{TL}(\neg, \supset)$ and $(\psi \supset \theta) \in \mathcal{L}_{TL}(\neg, \supset)$.

Given the restricted languages, classical two-valued semantics is correct and complete for each of these logics: it suffices to evaluate formulas of the restricted language – $\mathcal{L}_{GL}(\neg, \supset)$, $\mathcal{L}_{KL}(\neg, \supset)$, or $\mathcal{L}_{TL}(\neg, \supset)$ – assigning values to these formulas in the full language ($\mathcal{L}(\neg, \supset)$).

Another solution is simply to eliminate all cases of vacuity, both those cases where an atomic formula cannot assume the designated value, and those cases in which a molecular formula cannot assume the designated value or a set of formulas cannot assume simultaneously the designated value. But this, perhaps, is an excessively high price to pay and perhaps even comparable to eradicating a disease at the expense of the patient's life!

References

Aristotle (1938), Prior analytics, *in* J. Henderson, ed., 'Categories. On Interpretation. Prior Analytics', number 32 *in* 'Loeb Classical Library', Harvard University Press, pp. 181–531. Translated by Hugh Tredennick.

Bochvar, D. A. (1981), 'On a three-valued logical calculus and its application to the analysis of the paradoxes of the classical extended functional calculus', *History and Philosophy of Logic* 2, 87–112. Translated by Merrie Bergmann.

da Fonseca, F. F. (2007), Function and nature of infinite judgements: predicative aspects of negation in the critique of pure reason (in portuguese), Dissertação de Mestrado, Pontifícia Universidade Católica do Rio de Janeiro.

Gentzen, G. (1936), 'Die Widerspruchsfreiheit der reinen Zahlentheorie', *Mathematische Annalen.*

Glivenko, V. (1929), 'Sur quelques points de la logique de m. brouwer', *Académie Royale de Belgique - Bulletins de la classe des sciences.*

Kolmogorov, A. N. (1967), *From Frege to Gödel: a source book in mathematical logic, 1879 - 1931*, Harvard University Press, chapter On the principle of excluded middle (1925), pp. 414–437.

Paiva, V. & Pereira, L. C. P. D. (2006), 'A short note on intuitionistic propositional logic with multiple conclusions', *Manuscrito* 28, 317–329.

Pereira, L. C. P. D., Haeusler, E. H. & Medeiros, M. P. (2008), 'Some results on fragments of classical logic with negation (in portuguese)', *O Que Nos Faz Pensar* 23, 105–111.

In a wittgensteinian light

Gisele Dalva Secco*

* Departamento de Filosofia
Universidade Federal do Rio Grande do Sul
gisele.secco@ufrgs.br

It is not an exaggeration to say that philosophy and mathematics are in a kind of atavistic relationship with each other — at least as long as they have been practiced and conceived since Ancient Greece. From Plato to Wittgenstein we can find in the multifaceted universe of philosophy comparative considerations between these two forms of human knowledge. In most cases these comparisons point to differing aspects of the activities carried out by philosophers and mathematicians. Although the evaluation of these differences varies a lot, the variation in the evaluation of those differences is indexed by the election of one of them (mathematics) as a model to be observed by the other (philosophy).

Mathematical proofs are the central point of this story of comparisons and evaluations since they often serve as the standard procedure to authenticate knowledge. Philosophers sometimes thought philosophical practice was supposed to imitate (e.g. Spinoza) and sometimes to be opposed to mathematics (Kant). In any case it seems not to be necessary to spend much time trying to justify the philosophical interest in those aspects of mathematical proofs that show their epistemic strength, since they obviously deserve to be part of any investigation that aims at scrutinizing the different manifestations of human knowledge as well as their connexions.

Now, it isn't any novelty to state the importance of mathematics to Wittgenstein's philosophy, nor to recognize the importance of his philosophy in the twentieth century. What seems to be in the order of the day, after a long tradition of misunderstandings in what concerns Wittgenstein's philosophy of mathematics, is to show some of the ways through which his philosophical explorations are pertinent to give an account of actual mathematical practices. Therefore, the aim of this paper is to delineate an argument in favour of the idea that at least some points in Wittgenstein's philosophy of mathematics are especially appropriate nowadays by an example: a discussion about the philosophical controversies around the notorious proof of the Four-Colour Theorem.

1   The Four-Colour Theorem's proof and some philosophical consequences

When the proof of the Four-Colour Theorem (4CT) was published in a pair of papers written by Kenneth et al. (1977); Appel et al. (1977) a history of 125 years was completed — and yet another had begun. The mathematical history, dealing above all with topics in topology, combinatorics and graph theory, had initiated with a mathematics teacher's curiosity about the possibility of colouring any map with just four colours in such a way that no region that share a border is given the same colour. This curiosity received the status of a mathematical problem by means of Augustus De Morgan, who in a letter to Hamilton presented it as *a fact that he did not know neither if it was a fact*[1]. We find in his letter a constant aspect of that mathematical history: asking about the possibility of producing a simple solution to a problem whose formulation is simple. In fact, as we'll see, the task turned out to be a "tricky work."[2]

The 4CT states, be it the vocabulary proper to the field of combinatorics or in that of the topology[3], that every map under certain circumstances (normal planar maps) can be coloured with just four colours without any adjacent region having the same colour. The proof is a *reductio ad absurdum* in which the initial assumption posits the existence of a map that demands five colours to be coloured and that allegedly is the smaller of such maps, i.e., a *minimal* five-chromatic normal map. The reductio includes a proof by cases: in applying the so-called reducibility methods one can construct a set U of configurations from which every five-chromatic map has to contain at least one (hence, an *inevitable* set). Hence, it is possible to show that the initial map can be smaller, not being the minimal map it was suppose to be. The construction of the *inevitable set of reducible configurations* that contradicts the initial assumption of the *reductio* is resolved with three cases, one of which requires more than a thousand subcases that are practically impossible to be constructed without the execution of computer programs.

The main reason why this proof provoked a certain commotion in the mathematical and in the philosophical communities is the indispens-

---

[1]"A student of mine asked me today to give him a reason for a fact which I did not know was a fact, and do not yet". (Fritsch & Fritsch, 1998, pp. 8)

[2]*Idem.*

[3]The topological version of the 4CT affirms: "For every map there exists an admissible 4-colouring" (Op. cit., p. 86) while in the combinatorial one, without any reference to geometry or topology: "Every planar graph has an admissible vertex 4-coloring." (Op. cit., p. 149)

able participation of computers in its construction. In what concerns the members of the mathematical community, the responses to Kenneth et al. (1977)' result are twofold. On the one hand, some criticism is driven to the alleged lack of structure (and elegance) of the proof — an aspect directly related to the combinatorial nature of the solution. On the other hand, graph theorists claim, since the first critiques, that the kind of computational complexity of the 4CT's proof isn't different from the complexity of any proof in this mathematical field[4]. In the first group were to be found those who couldn't accept a computer assisted proof even if it had some structure or elegance, while in the second group were the defenders of the use of computers in the universe of mathematical practices without any fear related to the urban legend about a hidden error in the programs.

Although Kreisel (1977) and Wang (1981) were the two first philosophers that publicly mentioned the 4CT's proof, its philosophical citizenship was acquired through the paper "The four-color problem and its philosophical significance" (Tymoczko, 1979). In this paper, Tymoczko elaborates what I call the *introduction of the experimentation in mathematics via 4CT argument* (IEA). IEA's starting point is the observation that 4CT's proof is neither producible nor verifiable by one person in a lifetime. The argument then associates this fact with the use of computers, as if they were instruments of the same kind used in scientific experiments, to conclude for the necessity of revision in the standard conception of mathematical proof in such a way that it can include the possibility of the same kind of error we can find in experimental procedures in mathematics. Our reconstruction of Tymoczko's IEA is the following:

α The major characteristics of mathematical proofs, traditionally considered as "a priori deduction of a statement from premises" (Tymoczko, 1979, pp. 58), are these: (α_a) proofs convince, (α_b) proofs are surveyable e (α_c) proofs are formalizable;

β The 4CT's proof, although having (α_a) and (α_c), isn't (α_b), since the calculi executed with computational aid cannot be verified by a person in a lifetime;

---

[4]According to Swart (1980) they can be divided into three parts: "(i) Establishing the fact that the theorem is true provided a certain set of graphs, configurations, or — in general — cases possesses (or do not posses, as the case may be) a stated property. (ii) Obtaining an exhaustive listing of these cases. (iii) Confirming that all the members of this set do posses the required property. The finite set of cases concerned may, at one extreme, be so small and so simple that the case testing can be done in our heads, or it may, at the other extreme, be so large and/or so complicated that it is impossible to carry out without the help of a computer." (p. 699)

γ The use of computers in mathematical proofs introduces experimentation in the mathematical domain given that the reliability in the machines rests "on the assessment of a complex set of empirical factors" (Tymoczko, 1979, pp. 74); furthermore, in the case of 4CT, one of the uses of the machine is combined with the introduction of probabilistic reasoning;

δ Hence, by appealing irrevocably to the execution of computer programs, 4CT's proof "makes 4CT the first mathematical proposition known a posteriori", what forces us to change "the sense of the concept of âĂŸproof'" (Tymoczko, 1979, pp. 58). Therefore, mathematics is, once and for all, subject to the same kind of error typical of experimental methodologies.

It is not my aim here to give an analysis of the IEA. However, I would like to stress that the notion of surveyability is a key notion in Tymoczko's IEA: it is a missing aspect of 4CT's proof since "no mathematician has seen a proof of the 4CT, nor has any seen a proof that it has a proof." (Tymoczko, 1979, pp. 58) At least in a first argumentative move, the author identifies the surveyability of proofs with the capacity to be visualized, in such a way that Appel and Haken's work could be considered as a counterexample to the "traditional" concept of proof for which surveyability is a central characteristic. But what does it mean to say that no mathematician saw 4CT's proof?

Indeed, Tymoczko uses all the expressions corresponding to see, verify and survey in a very flexible way. Sometimes he seems to accept that to survey corresponds to a general inspection — in the sense of understanding the logical articulations of the main concepts activated in the proof, its "overall logics" — while at times he took the surveyability as the mark of the capacity to realize an algorithmic check. He still identifies this capacity with the possibility of the manual realization of the calculi involved in the proof of the key lemma. Even with this kind of semantic fluctuation in the IEA the author tenaciously insists in the impossibility of the step-by-step verification once "no computer has printed out the complete proof of the reducibility lemma" (Tymoczko, 1979, pp. 68). In fact, the immediate recognition of the fact that "such a printout [wouldn't] be of much use to human mathematicians" (loc. cit.) seems to indicate that the second sense of surveyability doesn't contribute for the first one. It is worth to notice that the distinction between local and global surveyability, suggested by Bassler (2006), could be of a good use here. Local surveyability corresponds to the capacity of checking the proof step by step and global surveyability is related to the possibility of grasping the sufficiency of the steps for proving the theorem. What is important to 4CT's case in Bassler's distinction, as we well see, is that he claims that global surveyability doesn't require local surveyability.

Another thing that seems worth to notice is that the argument revisits some traditional problems in the history of philosophy, like the distinction between *a priori* and *a posteriori* (knowledge and propositions), the question of the surveyability of proof procedures (that can be traced back to the Cartesian discussions about the possibility of grasping long chains of reasoning *d'un seul coup d'oeil*) and the topic on the role of errors in the experimental sciences by contrast with the supposed absence of them in the mathematical domain. Tymoczko's interests in such classical topics are made clear when he states the more general consequences of his argument, in terms of the urge for abandoning or modifying "many commonly held beliefs about mathematics" such as:

1. All mathematical theorems are known a priori.

2. Mathematics, as opposed to natural science, has no empirical content.

3. Mathematics, as opposed to natural science, relies only on proofs, whereas natural science makes use of experiments.

4. Mathematical theorems are certain to a degree that no theorem of natural science can match. (Tymoczko, 1979, pp. 63)

Philosophical rejoinders to the consequences of Tymoczko's argument and to the argument itself — the history that had begun whit the end of the mathematical one, the proof of 4CT — vary in a great deal of aspects. They vary from the degree of attention to the details of Tymoczko's presentation of 4CT's proof all the way to the amount of historical and argumentative acuity of the paper, passing through the evaluation of how good informed he was about the development of the computer science at his time, until the recognition of the weak grounding of his "quasi-empirical" standpoint. At any rate, the debates around 4CT's proof induced by Tymoczko's paper doesn't have the nature of a *disputatio* and some responses are elaborated until nowadays.[5]

In one relatively recent paper about the connexions between computer assisted proofs and proofs about computer programs Prawitz (2008) provides a partial reconstruction of the scenario in which the controversies

---

[5]Cf. McEvoy (2013).

about Appel and Haken's result were developed. His panoramic view is composed in such a way that we can identify two major groups in dispute: those who affirm and those who deny the occurrence of any significant change in mathematical practices since the 4CT's proof. Tymoczko, of course, is located in the first group. The second group, on the other hand, is divided in two smaller groups: one composed of those who don't accept that allegedly empirical procedures — such as the use of computers in the proof of the 4CT — have any meaningful role in proofs (Teller, 1980) and another composed of those who believe that those procedures don't constitute any such novelty in mathematics (Detlefsen & Luker, 1980). For these authors, empirical aspects in calculating procedures are normal and there is no problem in partially accepting empirical grounds for proofs, especially for such heavily computational proofs (in the sense of the reliability of an enormous amount of calculations to be concluded) like 4CT's proof.

It is not my aim here to reconstruct the narrative à la Prawitz about these controversies, not even criticize Prawitz's solution[6] — that, in the end, sustains Tymoczko's point about the empirical grounding for the 4CT. My aim here is to point out one aspect of Prawitz's argumentation strategy that can serve as a bridge to an alternative way to connect the philosophical discussions about the 4CT's proof with some wittgensteinian topics in philosophy of mathematics. Prawitz offers an analysis of some conceptual issues about proofs suggesting that it is more important to focus our attention on determining what it is to have proved something, i. e., what it is to be in possession of a proof of a certain proposition, than to ask what are proofs. In line with his constructivist heritage he affirms that we are in possession of a proof procedure when we carry out an operation on given grounds, transforming them in the grounds for the conclusion of a deductive inference.

Prawitz offers an analysis of some conceptual issues about proofs suggesting that it is more important to focus our attention on determining what it is to have proved something, i. e., what it is to be in possession of a proof of a certain proposition, than to ask what are proofs. In line with his constructivist heritage he affirms that we are in possession of a proof procedure when we carry out an operation on given grounds, transforming them in the grounds for the conclusion of a deductive inference.

One way to understand Prawitz's strategy is to activate the distinction between proof-act, proof-object and proof-trace suggested, in the context of a discussion related to proof theoretical semantics, by (Sundholm et al.,

---

[6]This reconstruction and the outline of a criticism on Prawitz's perspective are given in the fifth chapter of my PhD thesis (Secco, 2013).

1993). One advantage of this distinction consist in escaping the usual one between process and product once it introduces the elements by which we pass from one (the act of proving) to the other (the proof as an object produced by the act of proving) — the traces or instructions that regulate the process. In this sense we could think that Prawitz is claiming that proof-acts require the capacity of carrying out an operation that transforms given grounds for the premises into a ground for the conclusion while proof-objects are accessible only for those who can perform the act (something like: we can only have a complete map of a path we can walk through). Hence the importance of Prawitz's question about "what kind of things can amount to a conclusive ground for an assertion" Prawitz (2008, pp. 84) by contrast with the question "What is a proof?" that inevitably induces to think about proofs in terms of objects and, by its turn, to ignore the "performative" aspect involved in the idea of a proof as a transformation of grounds.

Consequently, from Prawitz perspective, although it could be justifiable to affirm the possibility *in principle* of giving a "traditional" (deductive) proof of the 4CT's (or, better yet, the proof of the reducibility lemma, that demands the workings of the computer), the justification we now have is not a deductive but an empirical one. For Prawitz, even if it can be proved deductively *that* "the program used by the computer to derive the key lemma was correct, we cannot prove deductively that the computer executed the program correctly." Prawitz (2008, pp. 91) According to the reading via the distinction between *proofs as acts*, *proofs as objects* and *proof as traces* Prawitz is suggesting that we can deductively prove that the program understood as a proof-object is correct — hence we have *a priori* grounds for the assertion according to which the object analyzed is indeed a proof — although the program understood as a proof-act cannot be deductively proved correct. "That the computer executed the program correctly" is, according to Prawitz, an assertion with a partial empirical ground.

Now I said before that Prawitz end up his analysis sustaining the same Tymoczko's point about the empirical grounding for the 4CT. To be fair, the similarities between Tymoczko's and Prawitz's conclusions can be stated only if we introduce some nuances while considering the arguments of each one of them. In this sense it is imperative to recognise the sophistication of Prawitz's approach, since he never proposes any kind of revision in the "traditional" concept of proof nor claims, like Tymoczko does implicitly in (γ) and (δ), that mathematics had almost turned into an empirical science given the possibility of error lurking the proof of the key lemma of 4CT's proof. On the contrary, even though he agrees with Tymoczko that

> If a theorem has been established only by relying on com-
> puters as in the proof of the four-colour theorem or in a proof
> that involves big computations, then the proof is not entirely
> deductive, and there is the undeniably epistemological conse-
> quence that the theorem is known only a posteriori. Prawitz
> (2008, pp. 92)

Prawitz distinguishes between this epistemological consequence and
the question about the (supposed lack of) confidence in the partially em-
pirically grounded deductive procedures. He affirms that it is a common
occurrence to find ourselves in a situation where "a computer report of
having found a specific proof may rightly be deemed as very trustworthy
and as more trustworthy than a corresponding report from a human of
having found a deductive proof" Prawitz (2008, pp. 92-3)

2 Between proofs and experiments: introducing wittgen steinian topics

While Tymoczko sustained that Appel and Haken's result was a kind of
hybrid between proof and experiment in virtue of the "calculatorial gap"
that was filled by an unsurveyable experiment (the running of the calculi
in computers) the first rejoinders attitude towards the question about the
philosophical significance of 4CT's proof was negative. Teller (1980), for
example, sustains that even if there are some empirical aspects in the proof
they do not play any proper role in it. Detlefsen & Luker, on the other way,
claim that the from a certain wittgensteinian perspective both of their positions can be evaluated in a way that the presence of such empirical aspects
is not any novelty in mathematical practices. In no of these rejoinders we
find a clear approach to the question of the surveyability of proofs (and
experiments or even calculations). It is only when a wittgensteinian perspective entered the scene that the notion of surveyability begin to receive
the proper attention given its role in the IEA.

Although agreeing with some premises of the IEA, Shanker (1987) proposes a much more drastic conclusion. Proofs and experiments, he claims
in a seemingly wittgensteinian way, are categorically distinct procedures.
Thus, sustaining the introduction of experiments in mathematics would be
nothing more than a category mistake. The wittgensteinian razor of this
author ends up suppressing both the procedure (the 4CT's proof) and its

result (the 4CT) from the domain of mathematical practices once it is, so claims Shanker, an experiment.

It is as if once we delegated to computers the calculating tasks involved in 4CT's proofs we had, so to speak, lost the access to the normative liaisons at stake in it. Shanker exemplifies this kind of relation referring to the tractarian concern with "the relation of a number to the law which generates the series in which it occurs. It is the law governing the expansion of the series, not the actual expansion of the series, what must be surveyable." (Shanker, 1987, pp. 128) As it is known the early Wittgenstein wanted to construct an intensional conception of mathematics through which the totality of numbers would not be understood as objects but instead as properties of a formal series. But how could this example be associated with the 4CT's proof? Shanker doesn't give a direct answer but introduces another example (extracted from the so called intermediate period of Wittgenstein's thought[7]):

> Wittgenstein argued that the colour octahedron must be surveyable in the sense that the logical articulations forged by the grammatical construction are perspicuous. Likewise, a proof must be surveyable in the sense that we can grasp the 'law' forged by the proof: 'I must be able to write down a part of the series, in such a way that you can recognise the law. That is to say, no description is to occur in what is written down, everything must be represented' (PR §190). But this is precisely the condition which the Appel-Haken solution fails to meet: what we are given just is a description of U — together with the operations which the computer has performed to test its reducibility — rather than a 'manifestation of the law' for the generation of the unavoidable set of reducible configurations. (Shanker, 1987, pp. 153)

A mathematical proof is then surveyable when the logical, conceptual or normative relations between its steps are graspable in the same way "the logical articulations forged by the grammatical construction" in the wittgensteinian model of surveyability: the colour octahedron. Here the requirement according to which no description can occur into a proof is clearly used by Shanker to reinforce the thesis according to which the 4CT's proof doesn't fit with Wittgenstein's considerations about the surveyability of proofs.

---

[7]The period goes from 1929 — when returning to Cambridge Wittgenstein also returns to philosophical activities — to 1934, although some texts from 1936 are counted as works from this period.

One immediate criticism against Shanker's introduction of wittgen-steinian themes in the discussion about the 4CT's proof was that he erected his argument against the mathematical nature of the Appel and Haken's procedure (and result) based on texts from the most chameleonic period of Wittgenstein's thought — and above all from texts whose central ideas, like that of a rule of grammar, are greatly modified in Wittgenstein's last writings. I am not claiming that a careful analysis of these texts is irrelevant for anyone willing to understand Wittgenstein's philosophical development[8].

However Shanker is right when he affirms that there are descriptions in the presentation of 4CT's proof. These descriptions occur in the second paper presenting the proof (Appel et al., 1977) and they refer to the programs executed by computers in the construction of the cases needed for the proof of the key reducibility lemma. Nevertheless maybe the occurrence of these descriptions isn't a sufficient reason for the exclusion of the procedure from the normative domain of mathematical practices. And the elements for justifying this hypothesis seems to be available in Shanker's own text, not to say in Wittgenstein's remarks themselves.

Let's return for a moment to Wittgenstein's early interest in the sense according to which we can talk about the totality of numbers mentioned by Shanker. The philosopher intended to clarify some confusion that can arise when one identifies finite totalities with infinite processes or series ("formal series" in the sense of the applicability of an operation). The point is that we can (at least in principle) construct all numbers throughout the successor operation, the "law" generating the infinite series of the numbers. For Wittgenstein, this is the only legitimate sense in which we can talk about the infinite totality of the numbers.

---

[8]My point is to stress that it isn't possible to sustain any sensible wittgensteinian stand-point about mathematical proofs, and inevitably about his strong case with the distinction between proofs and experiments, considering only middle period texts. If there is something like Wittgenstein's "final words" about this issues they can be found in the late period — and by late I mean not only the *Remarks on the foundations of mathematics* (RFM), notes organized by Von Wright, Rhees & Anscombe (1978), from 1937 to 1944, and the *Lectures* on the same theme (LFM), given in 1939 an edited by Diamond (1976), but also his ultimate notes, *On certainty* (Wittgenstein, 1968). Assuming this is a good point, Shanker's seemingly wittgensteinian argument is not so wittgensteinian as he would like and also do not serve as a good clarifying key for the philosophical considerations on 4CT's proof. Of course someone could argue that the interesting thing here is not to construe the most legitimate wittgensteinian reading of Appel and Haken's work but to scrutinize the possible correction of Wittgenstein's point of view, be it from the intermediate or from the late Wittgenstein. To this rejoinder I would respond that exactly because of the highly changing aspects of his thought in the middle period the safest route seems to be to look to the later texts in our search for more stable positions — those texts, as I would like to show, are the more adequate ones if we are to clarify some points on the debate surrounding the 4CT's proof.

So, why couldn't we say — applying this idea to the case of 4CT's proof — that the descriptions of the operations performed by the computer to test the reducibility of the set U are exactly the "manifestation of the law" for the generation of the unavoidable set of reducible configurations Shanker is demanding in his wittgensteinian vein? In proof-theoretical semantics vocabulary we could say that what must be surveyable here is the description of the effective procedure to execute an operation, not the execution of each and every one of it. Activating the distinction between local and global surveyability mentioned above we could also say that althought he 4CT's proof is in part *locally unsurveyable* (for humans but not for the machines) it is *globally surveyable*. And this is so not only because of what can be called the *functional aspect* of the descriptions occurring into the proof, i.e., the fact that they can *work* as a prescription to execute the calculus operations realized by the running of computer programs but also because, as shown from Prawitz's perspective presented before, we can deductively prove that the programs are correct. Or, what amounts to the same, we have control over "the laws" for the generation of U. Anyone inclined to implement the calculi by hand could do it, albeit with the ungrateful result that it would be an impossible task to finish in a lifetime. In any case the person in question, just as the machine, couldn't be regarded as realizing an experiment, but as calculating.[9]

In order to conclude the sketched panorama over the field of disputes around 4CT's proof it would be good to show how the "characteristic Wittgensteinian invention" (Mühlhölzer, 2006), the distinction between proofs and experiments, was mobilized from Shanker's approach onwards. Stillwell (1992), for example, claims that the distinction has to be better understood before being thrown into the controversies about the 4CT's proof. For Stillwell those disputes — specially Detlefsen and Luker's response to Tymoczko's IEA — raised some threats to Wittgenstein's apriorism and it is in the context of this problem that she develops some minor considerations on Appel and Haken's work. Aiming to defend the pertinence of the dichotomy proof versus experiment for her discussion the author lists a "network of logical points of contrast between proofs and experiments" (Stillwell, 1992, pp. 79), which can be present as follows (where **Pn** corresponds to logical or conceptual aspects of proofs and **En** to logical or

---

[9]It is not worthless to remember that from the *Tractatus* onwards Wittgenstein always insisted in this point. "Calculation is not an experiment", states the final phrase of a couple of tractarian aphorisms (6.233 and 6.2331) dealing with the necessity of some kind of intuition for the resolution of mathematical problems, to which Wittgenstein responds that it is the language itself, manipulated in the process of calculating, that brings about this intuition.

conceptual aspects of experiments):

P1 Surpass de descriptive domain of language: they introduce new concepts and do not mobilize propositions proper.

P2 Repeating a proof implies the necessary repetition of the result.

P3 Its realization don't depend on material conditions.

P4 Can be realized imaginatively.

P5 Errors nullify the procedure.

E1 Can be completely describable (and understandable) in empirical terms, from propositions proper.

E2 Repeating an experiment implies the possibility of different results.

E3 Its realizations depend on precise material conditions.

E4 Cannot be realized imaginatively.

E5 Errors do not nullify the procedure.

Of course a careful comparative analysis of each pair of characteristics (**P1** and **E1** etc.) is needed for a complete account of Stillwell's defense of Wittgenstein's apriorism with respect to "the content of our demonstrative knowledge" (Stillwell, 1992, pp. 74). However, what I want to point out here is something much more simple that the execution of this task[10]. It has to do with a possible complementation of Stillwell's argumentative moves, which also gives me a chance to return to Prawitz's solution for the 4CT's proof case. Let me present telegraphically that movement and then delineate the referred complementation.

The main concern Stillwell has with Detlefsen and Luker's account of 4CT's proof is related to their allegation that every calculus or proof episode contains an eliminable empirical element, introducing in the discussion something she wants to avoid — an epistemological aspect:

> Detlefsen and Luker are suggesting that a (rational) calculator cannot become convinced that a construction C establishes result R unless she/he believes both that no arithmetical errors occur in C and that, in the pertinent language, the result obtained or found in C is "R". [...] Accordingly, Detlefsen and

---

[10]In fact, this task was carried out in the sixth chapter of Secco (2013).

Luker seem right to say that the belief that a specific construction is a proof — if this species of belief is possible — rests on empirical beliefs. In that event, they also would be right that self-sufficient proofs — if proofs be specific constructions — must have empirical content. (*loc. cit.*)

Stillwell concedes to her opponents that Wittgenstein wouldn't identify the conviction that a construction is a proof with beliefs about specific constructions — what would be contrary to **P3**. Both approaches, Stillwell's and Detlefsen and Luker's could then be harmonized in regard to the idea that proofs "need nothing outside of itself to be convincing" in Tymoczko's expression.

It is known that since the *Tractatus* Wittgenstein credited autonomy from reality to mathematical "propositions" and procedures, in the sense of having no descriptive content like genuine propositions. The problem for Stillwell is that even using the idea of autonomy of proofs, Detlefsen and Luker maintain the identification between calculus or highly calculatorial proofs and experiments. Hence, the authors generate an insuperable wittgensteinian tension: in what sense an extremely computational proof couldn't depend on descriptive content or empirical conditions and at the same time be an experimental procedure?

To respond adequately this question the author would have to deal with the notion of calculus or computation worked out by Detlefsen and Luker, which can be approximated to the heritage of symbolic conceptions of knowledge and mathematics — according to which to calculate is to manipulate signs in symbolic structures, understood as "systems of physical objects subjected to operations of construction and transformation according to rules" (Esquisabel, 2012, pp. 21) and in which we can also find an elegant articulation between local and global surveyability[11]. At any rate, Stillwell prefers to focus on the empirical nature of the allegation that the computer in fact did what it was supposed to do (to execute correctly the calculations programed by Appel, Haken and Koch) to affirm that Wittgenstein sustained a "dual" point of view on proofs: in one hand, proofs are practical and variable phenomena (with empirical aspects, what could afford the empirical allegation just referred) while in the other they have the force of a standard, being independent of "particular configurations":

---

[11] Between those functions one finds a computational (which can me identified with local surveyability) and a cognitive function (related to the idea that in this kind of thought we grasp in the syntax of a symbolic system the structures being worked with. This grasping can be associated with global surveyability. Any resemblance with the aphorisms cited in note 9 isn't mere coincidence).

> To summarize, Wittgenstein holds that understanding a proof
> in a sense "transcends" our grasp of specific constructions, for
> our conviction that such and such is a way of proving X is not
> identifiable with, and does not involve, belief about spatial or
> temporal constructions. Wittgenstein also holds the stronger
> view that knowledge of proof is independent to a significant de-
> gree of beliefs about particular configurations. Still, it is not as
> if one could learn a proof with no appeal whatever to the latter.
> In addition, some of the above citations may themselves seem
> to intimate that "grasping a proof" at least indirectly includes a
> reference to our knowledge of tokens. (Stillwell, 1992, pp. 78)

What I would like to propose as a complementation, if not a correction,
of Stillwell's approach is the following. When appealing to the process of
learning a proof to stress that Wittgenstein recognizes them as acts depending on "particular configurations" Stillwell has made a positive step
forward if compared to Shanker's approach because she is pointing to the
growing attention the philosopher gave to mathematics as a human praxis,
a knowledge we learn from each other, being trained in the use of mathematical symbols for operations, constructions of "new (conceptual) roads"
etc. The problem seems to be that even highlighting the anthropological
aspect of mathematical practices Stillwell forgets one aspect the philosopher included between the notes of the concept of Übersichtlichkeit, which
in the end is the decisive criterion for the distinction proof versus experiment (she even refuses to talk much about surveyability or perspicuity,
preferring to use "holism" to refer to **P2**): the easiness with which a proof
is reproduced. In the first paragraph of the third section of RFM Wittgenstein says:

> 'A mathematical proof must be perspicuous'. Only a struc-
> ture whose reproduction is an easy task is called a "proof". It
> must be possible to decide with certainty whether we really have
> the same proof twice over, or not. The proof must be a con-
> figuration whose exact reproduction can be certain. Or again:
> we must be sure we can exactly reproduce what is essential to
> the proof. It may for example be written down in two different
> handwritings or colours. What goes to make the reproduction
> of a proof is not anything like an exact reproduction of a shade
> of colour or a handwriting.
> It must be easy to write down exactly the same proof again.
> This is where a written proof has an advantage over a drawing.
> The essentials of the latter have often been misunderstood. The

drawing of an Euclidean proof may be inexact, in the sense that the straight lines are not straight, the segments of circles not exactly circular, etc. etc. and at the same time the drawing is still an exact proof; and from this it can be seen that this drawing does not — e.g. — demonstrate that such a construction results in a polygon with five equal sides; that what it proves is a proposition of geometry, nor one about the properties of paper, compass, ruler and pencil. [Connects with: proof a picture of an experiment.] (Wittgenstein, 1978, pp. 143)

Hence, Stillwell put in motion a double strategy of disregarding. On the one hand, she didn't recognize a repeatedly emphasized aspect of the surveyability, that is, the *easiness of the reproduction* of a surveyable proof;[12] on the other hand, she deflated the importance of Wittgenstein's emphasis on seeing connections as something fundamental for proofs when stating, for example, that "the time has come to de-emphasize the visual elements in his account" (Stillwell, 1992, pp. 79) Now this deflation could serve as a point against Tymoczko, who made a confusion between the 4CT's proof not being completely visualized (in the sense of not being locally surveyable) and being partially empirical, but not as a good point in interpreting Wittgenstein's discussions on surveyability as a distinctive criterion of proofs and experiments. This is so because Wittgenstein's notion of visualization, the one at stake in his considerations about surveyability of proofs, doesn't correspond to seeing as a merely perceptive capacity but as the visualization of "what is essential" to the drawing or to the proof. And what is essential is the internal, logical or conceptual relations, what amounts to the same as saying that this kind of visualization corresponds to global surveyability, its symbolic aspect.[13] If so, Wittgenstein's perspective can contribute with some clarification for controversies on the 4CT's proof in a different way it has being done before. That is to say, in a way that not only recognizes the methodological or procedural mutations in mathematical practices (proofs, calculi etc.) but also intends to clarify the prose about it without any revisionist intention, aiming solely to describe

---

[12]Characterizing the surveyability as a purely formal or logical notion Mühlhölzer (2006) distinguishes the concept in four notions: (S1) Reproducibility; (S2) Easiness of (S1); (S3) Certainty about the identity of the proof; (S4) The reproducibility is the same as the reproducibility of pictures.

[13]Marion (2011) suggests the possibility of actualizing the interpretation of Wittgenstein's surveyability arguments in the light of the recent works on visualization in mathematics. My bet amounts to another possible liaison suggested by Stenlund (2013, 2014) between Wittgenstein and the symbolic tradition in mathematics, which can be successfully associated with the works presented in Lassalle Casanave (2012).

the processes through which those mutations constitute the working of the mathematical community.

For Wittgenstein it is only when mathematical proofs are taken as manifesting a descriptive use of language, or of the language understood as a system of representations and descriptions of some domain of objects, that it can be possible to make the mistake of considering a sentence like "Every planar map is four-colorable" as an empirical one. In the same sense it is possible (although wrong) to claim that the participation of computer machines in the execution of some calculi needed for proving a lemma the procedure recognized by the mathematical community (not without a tense period of persuasion, of course) makes the result not a mathematical but an empirical one. To reveal through conceptual elucidation the misunderstandings in the kind of prose that can lead to such relatively revisionist conclusions[14] was an important point for Wittgenstein:

> In a crude way — the crudest way possible — if I wanted to give the roughest hint to someone of the difference between an experiential proposition and a mathematical proposition which looks exactly like it, I'd say that we can always affix to the mathematical proposition a formula like "by definition".
> "The number of so-and-so's is equal to the number of so-and-so's": experiential or mathematical? One can affix to the mathematical proposition "by definition". This effects a categorial change. If you forget this, you get an entirely wrong impression of the whole procedure.
> The "by definition" always refers to a picture living in the archives there. — If we forget this we get into one queer trouble: one asks such a thing as what mathematics is about — and someone replies that it is about numbers. [...] I am trying to show in a very general way how the misunderstanding of supposing a mathematical proposition to be like an experiential proposition leads to the misunderstanding of supposing that a mathematical proposition is about scratches on the blackboard. (LFM, pp.

---

[14]If a philosopher claims that a decision of the mathematical community, like accepting Appel and Haken's procedure as a proof and not as an experiment, induces a conceptual change he is not being revisionist with respect to mathematics but to "our language" since, or so argues Tymoczko, we now have to use the concept of proof in a way that includes experimental elements, possibility of error etc. If the philosopher judges, like Shanker, that the mathematical community is wrongly applying the concept of proof to an experiment, he then is making exactly the kind of move I think Wittgenstein would not accept as a philosophical one: he is not worried with describing the decisions taken by mathematicians or with discussing the criteria they apply to the use of concepts like proofs or experiments, but legislating about the mathematical prose.

111-12)

Can't we say now that the 4CT is a proposition about nothing but the terms defined by the procedures involved in its proof? If the response is positive, it is because we can see the introduction of computers as a novelty in the methodological or constitutive domain of mathematical practices. In this sense the computer is part of the determination of the conventions from which 4CT's proof is constructed and then Prawitz's claim that the ground for the assertion of the key lemma is partially empirical doesn't pose a problem for a wittgensteinian reading of Appel and Haken's procedure and result, because we can understand the assertion that the computer did what is was supposed to do as being "in the archives there", *working into the deductive setting of the proof as an a priori proposition.*

Maybe it is worth to end by indicating that the idea of an originally empirical proposition being hardened into a rule[15], which can be seen as his specific type of conventionalism, isn't something exclusive of Wittgenstein's last observations (for examples, cf. OC §94-98, 308-309):

> Wittgenstein claims there [in the Yellow Book] that we can think that the same sentence undergoes "a transition between a hypothesis and a grammatical rule" (Wittgenstein 2001, 70). He also says in his lectures of 1934-5: "It is quite possible for a proposition of experience to become a rule of grammar" (Wittgenstein 2001, 160). In this case, the proof makes the empirical confirmation of the same sentence (hypothesis) superfluous. (Engelmann, 2009, pp. 109)

From this perspective, which can certainly be better presented and explored in future occasions, if the 4CT's proof introduces something new in the universe of mathematical practices it is nothing but a new technique for proving mathematical propositions, a technique including that sort of "methodological propositions" Wittgenstein talks about when dealing with cases of empirical regularities hardened into rules. Maybe following these hints, some of the problems indicated above could be solved. Still, in the philosophical domain we can say that the prose about 4CT's proof (produced not only by bad informed philosophers but also by mathematicians) introduced the necessity of the kind of clarification Wittgenstein was used to propose. It is highly probable that I didn't succeed in clarifying some of the topics that emerged in that prose as much as I intended. But at least the essay was done in a wittgensteinian light.[16]

---

[15]Cf. Steiner (2009) for a discussion on the topic.

[16]I would like to thank professor Luiz Carlos Pereira for the indications he gave me,

References

Appel, K., Haken, W., Koch, J. et al. (1977), 'Every planar map is four colorable. part ii: Reducibility', *Illinois Journal of Mathematics* **21**(3), 491–567.

Bassler, O. B. (2006), 'The surveyability of mathematical proof: A historical perspective', *Synthese* **148**(1), 99–133.

Detlefsen, M. & Luker, M. (1980), 'The four-color theorem and mathematical proof', *The Journal of Philosophy* pp. 803–820.

Diamond, C. (1976), *Wittgenstein's lectures on the foundations of mathematics Cambridge, 1939*.

Engelmann, M. (2009), 'The multiple complete systems conception as fil conducteur of wittgensteinâĂŹs philosophy of mathematics', **32**, 111–113.

Esquisabel, O. M. (2012), 'Representing and abstracting. an analysis of leibniz's concept of symbolic knowledge', *A. Casanave Lassale, Sutidies on logic, Symbolic Knowledge from Leibniz to Husserl* pp. 1–62.

Fritsch, R. & Fritsch, G. (1998), *Four-Color Theorem*, Springer.

Kenneth, A., Haken, W. et al. (1977), 'Every planar map is four colorable. part i: Discharging', *Illinois Journal of Mathematics* **21**(3), 429–490.

Kreisel, G. (1977), From foundations to science: justifying and unwinding proofs, *in* 'Recueil des travaux de l'Institut Mathématique, Nouvelle serié, Symposium: Set theory. Foundations of Mathematics', pp. 2–10.

Lassalle Casanave, A. (2012), 'Symbolic knowledge from leibniz to husserl'.

Marion, M. (2011), 'Wittgenstein on surveyability of proofs'.

McEvoy, M. (2013), 'Experimental mathematics, computers and the a priori', *Synthese* **190**(3), 397–412.

Mühlhölzer, F. (2006), '"a mathematical proof must be surveyable" what wittgenstein meant by this and what it implies', *Grazer Philosophische Studien* **71**(1), 57–86.

---

since early in my academic life, that thinking *with* Wittgenstein can be a good way to follow our own paths.

Prawitz, D. (2008), Proofs verifying programs and programs producing proofs: A conceptual analysis, in R. Lupacchini & G. Corsi, eds, 'Deduction, Computation, Experiment', Springer Milan, pp. 81–94. URL http://dx.doi.org/10.1007/978-88-470-0784-0_5.

Secco, G. D. (2013), Entre Provas e Experimentos. Uma leitura wittgensteiniana das controvérsias em torno da prova do Teorema das Quatro Cores, PhD thesis, Pontifícia Universidade Católica do Rio de Janeiro, Rio de Janeiro.

Shanker, S. (1987), *Wittgenstein and the Turning Point in the Philosophy of Mathematics*, State University of New York Press. URL http://books.google.com.br/books?id=wOR44vscOWEC.

Steiner, M. (2009), 'Empirical regularities in wittgenstein's philosophy of mathematicsâĂǎ', *Philosophia Mathematica* **17**(1), 1–34. DOI 10.1093/philmat/nkn016. URL http://philmat.oxfordjournals.org/content/17/1/1.abstract.

Stenlund, S. (2013), 'Wittgenstein and symbolic mathematics', *To appear.*

Stenlund, S. (2014), *The Origin of Symbolic Mathematics and the End of the Science of Quantity*, Vol. 59, B. Blackwell.

Stillwell, S. (1992), Empirical inquiry and proof, in M. Detlefsen, ed., 'Proof and Knowledge in Mathematics', Routledge, pp. 110–134.

Sundholm, B. et al. (1993), 'Questions of proof', *Manuscrito (Campinas)* **16**, 47.

Teller, P. (1980), 'Computer proof', *The Journal of Philosophy* pp. 797–803.

Tymoczko, T. (1979), 'The four-color problem and its philosophical significance', *The Journal of Philosophy* pp. 57–83.

Wang, H. (1981), *Popular lectures on mathematical logic*, Courier Dover Publications.

Wittgenstein, L. (1968), *On certainty*, Vol. 174.

Wittgenstein, L. (1978), *Remarks on the Foundations of Mathematics*, Vol. 7, B. Blackwell.

Wittgenstein on diagonalization

Guido Imaguire*

* Centro de Filosofia e Ciências Humanas
Universidade Federal do Rio de Janeiro
guido_imaguire@yahoo.com

Abstract

There are at least two points on which I totally agree with Luiz
Carlos Pereira: that winter is better than summer, and that Wittgenstein deserves much more attention than traditional logicians and
philosophers of mathematics usually pay to him. In this paper, I will
try to make sense of some of Wittgenstein's comments on transfinite
numbers, in particular his criticism of Cantor's diagonalization proof.
Many scholars have correctly argued that in most cases in the philosophy of mathematics Wittgenstein was not directly criticizing the
calculus itself, but rather the philosophical prose that goes along with
the calculus. But this does not mean that Wittgenstein has nothing
logically or mathematically substantial to say about the calculus itself.
Indeed, I think that in his criticism of Cantor's proof, Wittgenstein
provides an insightful logical argument.

1   Wittgenstein's Mysterious Passage

Wittgenstein changed his mind concerning many questions of the philosophy of mathematics from the *Tractatus* to his mature work. But his
critical attitude toward the existence and the status of infinite sets remains
consistent throughout. One fundamental idea in Wittgenstein's philosophy
of mathematics is that the word 'class' has two different meanings when
used in finite and infinite frameworks. When it is used in the context of
a finite framework, it means totality. But when it is used in the context
of an infinite framework, it means a rule-governed series, so to speak
'a process', or a mere possibility of a series constructed by means of an
operation (see Shanker, 1987, pp. 165). Thus, we should not interpret an
infinite sequence as already given in extension. Infinity is never more than
a mere possibility. Given that Wittgenstein maintained a skeptical attitude
concerning actual enumerable infinities, it is not surprising that he had an
even more critical attitude concerning non-denumerable infinities. In fact,

in the Remarks on the Foundations of Mathematics he qualifies Cantor's diagonalization proof as 'a puffed-up proof' (RFM II, §21) and as 'Hokus Pokus' (RFM II, §22).

Some interpreters have tried to make sense of Wittgenstein's criticism of Cantor. According to Shanker (1987, pp. 166), Wittgenstein's remarks on this point are extremely superficial and 'suffer from an acute lack of gravitas'. Further, Wittgenstein believed that he had perceived grammatical confusions in Cantor's transfinite theory. Marion (1998, pp. 200), probably because of his respect for such a well-established proof in the mathematical community, suggested a moderate interpretation, according to which 'Wittgenstein did not oppose the diagonal method itself, but only the interpretation of the result of its application in Cantor's proof'. I disagree with both, and think that Wittgenstein really has a substantial logical (not just 'grammatical', 'philosophical' or 'interpretational') point against the proof itself. I also disagree with da Silva (1993, pp. 95), who thinks that Wittgenstein's claim against Cantor lies in the method of *ad hoc* choices used by Cantor to define the diagonal number. If I correctly understand Wittgenstein, he saw a problem with the proof by diagonalization that even a partisan of infinitism should recognize.

This problem is expressed in an interesting although obscure passage of the Remarks (RFM II, §11):

> Da meine Zeichnung ja doch nur die Andeutung der Unendlichkeit ist, warum muss ich so zeichnen
> (Since my diagram is just an indication of infinity, why must I draw in this way:)

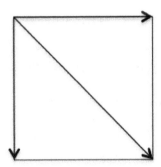

Und nicht so:
(Why not so:)

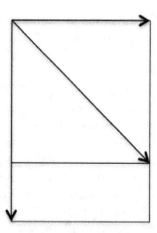

For facilitating reference, let us call the first diagram the 'A-diagram' and the second the 'B-diagram'. The only difference between them is the fact that the B diagram has an additional rectangle in the lower part. What is the meaning of this additional rectangle?

The quoted passage occurs at the beginning of the second part of the *Bemerkungen*, i.e., in the context of a discussion of Cantor's proof by diagonalization. As we all know, according to Cantor the set of real numbers in the interval between 0 and 1 is non-denumerable. Any enumeration of the set of real numbers in this interval would necessarily be incomplete. The explanation is as follows: given any list L of real numbers of the form

$0, \underline{a_1} a_2 a_3 a_4 a_5, \ldots$

$0, b_1 \underline{b_2} b_3 b_4 b_5 \ldots$

$0, c_1 c_2 \underline{c_3} c_4 c_5, \ldots$

$0, d_1 d_2 d_3 \underline{d_4} d_5 \ldots$

$0, e_1 e_2 e_3 e_4 \underline{e_5} \ldots$

. . . there will be a diagonal number (D), constructed with the diagonal decimals of this enumeration (the underlined digits), i.e., in this case, the real number: $0, a_1 b_2 c_3 d_4 e_5 \ldots$ Now, we simply define the anti-diagonal number (AD) as the number whose decimals are one-by-one different from the D, e.g., the number $0, a_1^\star b_2^\star c_3^\star d_4^\star e_5^\star \ldots$, such that $a_1 \neq a_1^\star, b_2 \neq b_2^\star, c_3 \neq c_3^\star, d_4 \neq d_4^\star \ldots$ Since AD differs at each decimal position from each number of L, it follows that AD differs from all numbers in L. The conclusion: L is incomplete, i.e., the set of real numbers is non-denumerable.

Most interpreters suggest that Wittgenstein's objection to Cantor concerns the non-constructivity of AD (see, e.g., Shanker, 1987; Rodych, 1999, pp. 195 and 281, respectively) or that the misleading aspect lies in talking of 'the set of all reals' or 'the real numbers in toto', since there is no given infinite totality. I agree with them, but I think Wittgenstein has more to say about this, and that this is exactly the meaning of the obscure passage.

To clarify Wittgenstein's mysterious quotation, the first thing we should notice is that the structure of the method of diagonalization could be represented as an A diagram. The left-to-right arrow of the A diagram represents the infinite sequence of decimals of the reals, and the top-down arrow represents the numerals we used to enumerate them. The diagonal arrow, of course, represents the D number.

My suggestion to interpret Wittgenstein's critical remark is as follows. Suppose that our task were to enumerate only the numbers with just 1 decimal. Our complete enumeration would simply be something like this:

1 − 0.0

2 − 0.1

3 − 0.2

. . .

10 − 0.9

It is, of course, impossible to enumerate the numbers with 1 decimal in 1 line — we would need exactly 10 lines to finish our task. Now, if our task were to enumerate all numbers with 2 decimals, i.e., from 0.00 to 0.99, by similar reasoning it is clear that we would need more than 2 lines — we would need exactly 100 lines. To enumerate the numbers with 3 decimals (0.000 to 0.999), we would need more than 3 lines — we would need exactly 1000 lines, and so on. More generally, in our decimal representation system (with digits $'0', '1', '2', \ldots, '9'$), for a set with n decimals, we will always need 10n lines to finish a complete enumeration. To put this in a negative way: It is generally impossible to get a one-to-one correspondence between the set of numbers with n decimals and a set with n members like that suggested in the following diagram:

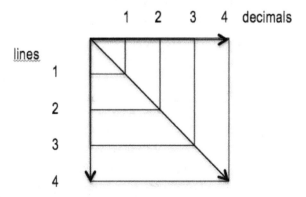

I think that Wittgenstein's point is exactly this: If, for any finite n, it is impossible to enumerate a set of numbers with n decimals in just n lines, why should we suppose that this method would work for an infinite n? If it is not possible to enumerate the set of all numbers with n decimals using n numerals, why should we be able to enumerate the set of numbers with ω decimals in ω lines? Even an infinitist should recognize this impossibility. In the form of a diagram: it is impossible to enumerate in this way:

n-decimals

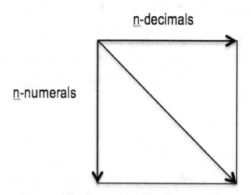

n-numerals

But, and this is the important point, it is still possible to enumerate the set of numbers with n decimals in $n + m$ lines (m is a finite number, in the case that n is finite, too), i.e., in this way

n-decimals

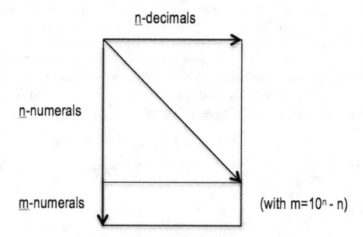

n-numerals

m-numerals   (with $m = 10^n - n$)

Cantor's conclusion was this: the AD number is missing, it is found nowhere in the list. Wittgenstein would probably reply: "Yes, you are right. In any such enumeration there is always a missing AD that is not in the rectangle in the A diagram — this is even the case with finite enumerations. If someone offers an enumeration of all numbers between 0.000 and 0.999 in only 3 lines, I can use diagonalization in order to create the missing number. But I can always add this number to the list — actually, it will appear in the m-part of the B diagram. Thus, not the A-diagram but the B-diagram is the most accurate representation of increasing finite or infinite enumerations. And in the B diagram there is neither a D nor an AD number."

There is an important connection between the point I am making here and the standard objection of the non-constructive nature of Cantor's proof. One of the main "insights" (or, if you will, "dogmas") of Wittgenstein's philosophy of mathematics is the gap between extensions and intensions that we mentioned in the introduction. As Wittgenstein put it, "the mistake in the set-theoretical approach consists time and again in treating laws and enumerations (lists) as essentially the same kind of thing" (Wittgenstein, 1974, pp. 461). Wittgenstein used the ironic expression "dots of laziness" in this context. We normally use these dots, as in "3.333333...," in order to express an extension that "could in principle be completed." Actually, this kind of expansion is not as bad as a lawless expansion, since a principle of construction could be offered for it. But sometimes we also use dots of laziness for lawless irrationals and lawless lists that are "free-choice" or irregular expansions generated by some non-mathematical means (Rodych, 1999, pp. 284). And the use of dots is not justified in these cases — it is not just a case of "laziness." Wittgenstein rejects both "lawless irrationals" and "lawless lists." And, as a matter of fact, both appear in Cantor's proof, wrongly represented by dots of laziness. Firstly, in each line of the proof, the dots (as in $0, a_1 a_2 a_3 a_4 a_5, \ldots$) represent an infinite lawless expansion. Secondly, in the vertical direction downward the dots represent the continuation of the randomly construed list. Thus, the AD number is non-constructive in a double sense. Contrary to this, in all the finite enumerations (from 0.0 to 0.9; or from 0.00 to 0.99; or from 0.000 to 0.999), the list and each number of the list is rule-governed, and the AD number is consequently constructive. But, and this is the important point, the "expansion" from the finite cases to the infinite cases is not just a difference in quantity, but in nature, for it represents a loss of regularity.

But is this really an argument against Cantor? It seems prima facie that Wittgenstein is simply affirming the impossibility of enumerating reals, just as Cantor did. But I think that behind this superficial convergence there is a deep disagreement. This should become clear when we discuss some possible reactions.

2 Possible Reactions

2.1 Notational System

One first possible — I think uninteresting — reaction to what has been shown here is this: Wittgenstein's argument is restricted to a particular system of notation, viz. the decimal system. And that is not logically necessary, but only contingent. We need 10^n lines in order to enumerate a set of numbers with n digits in the decimal system. But, in fact, Wittgenstein's argument applies to any system able to represent infinite totalities. Indeed, in his original 1981 paper Cantor considered the set of all infinite sequences of binary digits. He constructively showed that if $s_1, s_2, \ldots, s_n$ is any enumeration of elements from this set, then there is always an element of this set which corresponds to no s_n in the enumeration. To prove this, given as an enumeration of arbitrary members of this set the following:

$s_1 = \underline{0}, 0, 0, 0, 0, 0, \ldots$
$s_2 = 0, \underline{1}, 0, 1, 0, 1, \ldots$
$s_3 = 1, 0, \underline{0}, 1, 0, 0, \ldots$
$s_4 = 0, 0, 0, \underline{1}, 1, 1, \ldots$
$s_5 = 1, 0, 0, 0, \underline{0}, 1, \ldots$
$s_6 = 0, 0, 0, 0, 0, \underline{1}, \ldots$

he constructed the AD sequence, similar as before, and by replacing '1' with '0' and vice-versa got: AD = $1, 0, 1, 0, 1, 0 \ldots$ By construction, AD differs from each s_n, since their n^{th} digits differ.

But, of course, we can challenge this construction in the same way. Consider this binary system with 0 and 1. To enumerate instances (numbers or sets) with 1 digit, we would need 2 lines; to enumerate instances with 2 digits we would need 4 lines; to enumerate instances with 3 digits we would need 8 lines. Or, more generally, to enumerate instances with n digits we need 2^n lines. Again, the correct diagram is not A, but B — now with $m = 2^n - n$. To put this in a general form: the size of the additional part m of the diagram depends on the number of symbols in the system. The minimal possible notational system would be something like Hilbert stroke-numerals with just one sign 'I': I, II, III... But here the expression of non-denumerability is no more possible. Thus, to express non-denumerability we have to assume n^2. Note that, given n^2, since a real number is a number corresponding to an infinite sequence of the form $a, a_1 a_2 a_3 \ldots a_n$, with $n = \omega$, it follows that $m = n^\omega - \omega$. This means that m itself is non-denumerable! This leads us to the next possible reaction.

2.2 Is there any disagreement at all?

A second possible reaction is this. One could say that Wittgenstein's remarks do not represent an argument against Cantor's proof at all. For Cantor's proof was meant to show exactly the same result that I am attributing to Wittgenstein, viz. that an enumeration of reals is not possible. Indeed, Wittgenstein also recognized the utility of Cantor's proof, viz. to stop someone who tried to enumerate all real numbers (RFM II §13).

But, if I am interpreting Wittgenstein correctly, the point he made is this: Cantor was wrong both in his diagnosis and in his result. The proposed enumeration of reals is impossible not because the set of irrational numbers is non-denumerable, as Cantor supposed, but simply because an enumeration of numbers with n decimals in n lines is not possible, for any (finite or infinite) n. In the case of a finite n, by diagonalization we can create the missing AD number — but this number can and *will* appear in the m complement. Why should we assume something different for an infinite n? Of course, Cantor could reply: in the case $n = \omega$, the m complement is itself non-denumerable. So, with the B diagram you are simply establishing a one-to-one correspondence between two non-denumerable sets. It may be true, as Wittgenstein suggested, that we cannot enumerate a number with n decimals in n lines when n is finite. But that only shows that this method does not work for any finite n. It may still work for infinite numbers, i.e., when $n = \omega$. In fact, in this case m itself is non-denumerable.

I think Wittgenstein would not be impressed by this argument. It is important to note that Wittgenstein did not agree with standard intuitionism, which insists on the distinction between actual and potential infinities. For him, the word 'potential' or 'possibility' is misleading: the word "possibility" is of course misleading, since someone could say, 'Let what is possible now become actual' (PR §141). And this is exactly what happens in Cantor's proof: the infinite sequence, both the list of irrationals and the expansion of each irrational, is explicitly taken as given or 'completed'. When we think of mathematics as a procedure, as we really should, and not as a set of given results, we will note that in order to enumerate a number with n decimals in a system with s symbols, we will always need $s^n - n$ lines. The value of m increases exponentially with the value of n, but we can never effectively list all numbers with infinite decimals. In fact, we can never effectively complete the infinite expansion of each of the lawless irrationals. Here it becomes clear why dots of laziness are unjustified in diagonalization.

Now, just for the sake of argument, let us suppose that we can "effectively" and "actually" deal with infinite sets. Cantor must suppose this, for his proof only works if we assume that such an attempt to finish the enumeration was effectively and actually made. Remember that AD is only a definite number when all infinite digits are given. But, then, why should we not assume that the AD number does not occur in the complementary m part of the diagram? In the case of infinite sets, of course, the n will be infinite and enumerable (the left-to-right arrow and the top-down arrow in the A diagram). But if we are supposing that the work is 'really actually done', then we should be consistent and also assume that the complementary rectangle of the B diagram could also be finished — and AD will occur in it. For, why should we not assume this? Just because it is 'bigger', i.e., it is more than just enumerable? We cannot assume this, for this is just the point we still want to prove. To suppose that m cannot be finished because it is non-denumerable is a petitio principii. And since here $m = n^\omega - \omega$, m is 'long enough to go beyond' the n enumerated real numbers, covering inclusively AD — as we learned with a finite n. Thus, there is no missing number.

Shanker (1987, pp. 168) suggested that Wittgenstein was not undermining the existence of transfinite numbers, but rather only the interpretation according to which Cantor just 'extended' the domain of natural numbers. What Cantor really achieved is an extension of the concept of numbers. We could even think of the notion of number as a family-resemblance concept.[1] Indeed, I think that Wittgenstein could agree with Cantor in saying that transfinite numbers are no less 'real' than natural numbers. In fact, natural numbers and irrational numbers are similar in this point: they should be taken as a process, not as a given extension. But, contrary to Shanker, I think that in one aspect the case for reals is worse than the case for naturals. When mathematicians deal only with natural numbers in practice, they do not have to decide if the dots of the infinite expansion are somehow "given" in extension or not. This could remain an open (perhaps useless) philosophical question. But when they deal with transfinite numbers, as in the case of the diagonal method of proof, they have already decided that the dots are a representation of a given infinite extension — which means that they have already reached the wrong conclusion.

References

Cantor, G. (1891), 'über eine elementare frage der mannigfaltigkeitslehre', *Jahresbericht der Deutschen Mathematik-Vereinigung*.

---

[1]Indeed, in *Grundlagen* Cantor seems to assume only the idea that he was extending the notion of number, and not the domain of natural numbers.

da Silva, J. J. (1993), Wittgenstein on irrational numbers, *in* 'Wittgensteins Philosophie der Mathematik, Akten des 15. Internationalen Wittgenstein-Symposiums'.

Marion, M. (1998), *Wittgenstein, Finitism, and the Foundations of Mathematics*, Clarendon Press.

Rodych, V. (1999), 'Wittgenstein on irrationals and algorithmic decidibility', *Synthese* **118**, 279–304.

Shanker, S. (1987), *Wittgenstein and the Turning-Point in the Philosophy of Mathematics*, State University of New York Press.

Wittgenstein, L. (1974), *Philosophical Grammar*, Basil Blackwell. Anthony Kenny (trasn).

Wittgenstein, L. (1978), *Remarks on the Foundations of Mathematics*, Basil Blackwell.

Some remarks on proof-theoretic semantics and the universalist Perspective of language[‡]

Javier Legris*

* IIEP-BAIRES (UBA-CONICET)
Universidad de Buenos Aires
jlegris@retina.ar

In a famous paper from 1967, Jean van Heijenoort distinguished between *logic as calculus* and *logic as language* in order to describe two opposite trends in the earlier development of mathematical logic (see van Heijenoort, 1967b). The distinction was generalized by Jaakko Hintikka, who applied it on many occasions to the interpretation of 20th century philosophy (see for example the introduction of Hintikka, 1997). Originally, the distinction was related to the historical development of symbolic logic in the second half of the 19th century. Notwithstanding, it is also useful to show the different philosophical assumptions in current work in logic, as Hintikka has shown. According to the universalist conception of language, semantics cannot be defined in *our only language* without falling into a vicious circle. So, semantics cannot be expressible in the language. This fact motivated Hintikka to speak of the "ineffability of semantics". This paper is an attempt to discuss these two notions in relation to the proof-theoretic semantics, as it was characterized and carried out by Michael Dummett, Dag Prawitz and Peter Schroeder-Heister, among many others. The case of proof-theoretic semantics is quite interesting not only because it is an alternative to model-theoretic semantics, but also because of its roots in mathematical intuitionism. This school had its own conception about the role of language, as ordinary as formalized, in foundational issues. For the intuitionists language was secondary in the construction and justification

---

[‡]I would like to point out that the reflections included in this paper were inspired by talks given by Luiz Carlos Pereira in different issues of the *Southern Cone Coloquia of Philosophy of Formal Sciences*. I am grateful to Abel Lassalle Casanave for his valuable comments of an earlier draft of this paper and to Wagner Sanz for his suggestions that improved the paper. The paper was written as part of the exchange project CAFP-BA 042/12, between Argentina and Brazil and the research project PIP 11220080101334 from CONICET (Argentina).

of mathematics. Arendt Heyting introduced formalization *stricto sensu* in intuitionism, and therefore paved the way for proof-theoretical semantics for intuitionistic logical constants. In the paper Heyting's conception of formalization will be connected with the tradition of *symbolic knowledge* in formal sciences.

1 The Idea of a Proof-theoretic Semantics

As a result of the formal analysis of intuitionistic logic and the so-called BHK interpretation of logical constants, the idea of constructing a formal semantics based on the notion of proof took shape. It is known currently as *proof-theoretic semantics* being an alternative to modal-theoretic semantics, which prevailed (and still prevails) in symbolic logic. The term "proof-theoretic semantics" was proposed by Peter Schroeder-Heister at the end of the 20th century (for a comprehensive account on the subject, see Schroeder-Heister, 2012). According to the model-theoretic view, the meaning of a sentence is identified with its truth conditions. A consequence is logically valid if it is transmits truth from its premises to its conclusion, with respect to all interpretations. Thus, logical consequence can be established by showing that truth (in a model) is transmitted from the antecedents to the consequent of the consequence claim. Formal deduction systems are shown to be correct by proving that the consequences they generate are logically valid, and proof theory was looked upon as a part of syntax. Hence, model theory was seen as the adequate tool for semantics. The situation is quite different in proof-theoretic semantics. As Reinhardt Kahle and Peter Schroeder-Heister wrote some years ago:

> "Proof-theoretic semantics proceeds the other way round, assigning proofs or deductions an autonomous semantic role from the very onset, rather than explaining this role in terms of truth transmission. In proof-theoretic semantics, proofs are not merely treated as syntactic objects as in Hilbert's formalist philosophy of mathematics, but as entities in terms of which meaning and logical consequence can be explained." (Kahle & Schroeder-Heister 2006, pp. 503)

As Dag Prawitz pointed out, proof-theoretic semantics is not a *contradictio in adjecto*, because proof theory is not regarded as part of syntax (see Prawitz, 2006). The main ideas of proof-theoretic semantics proceed mainly from the system of natural deduction conceived by Gerhardt Gentzen around 1934 in his PhD thesis, but the BHK interpretation of logical constants is also in the background. Gentzen clearly stated in a now celebrated passage:

> "The introduction [rules] represent, as it were, the 'definitions' of the symbols concerned and the eliminations [rules] are no more, in the final analysis, that the consequences of this definition" (Gentzen, 1934, pp. 80)

A proper semantic treatment of this idea consists in connecting meaning with the notion of proof (at least in the case of logic and mathematics): the meaning of a sentence is determined by what counts as a proof of the statement (see v.g. Prawitz, 1998, pp. 44). Proof has an independent semantic value, unlike what happens in model-theoretic semantics, where proofs depend on truth transmission. Under these conditions, intuitionistic logic turns to be the privileged logic.

To some extent, the proof theoretic approach was supposed to result from reflections on the existing mathematical practice (see Prawitz, 1978, pp. 26). Hence, it would be a formal semantics closer to mathematical practice than classical semantics. Now, the interest in proof-theoretic semantic could go beyond logic and mathematics. As it was pointed out by Dag Prawitz, the proof-theoretic semantics is *verificacionist* in the sense that intuitionistic validity is based on considerations of meaning, rather than on ontological considerations (see Prawitz, 1998). Several attempts have been undertaken to elaborate this semantic program. For mathematical language it consists in the *intuitionistic* idea that the meaning of a sentence is given by specifying the form which a proof of the sentence should have, as happens with the BHK interpretation of logical constants.

2 Language as Calculus and Language as Universal Medium

In a now famous paper from 1967, Jean van Heijenoort distinguished between logic as calculus and logic as language in order to understand two opposite trends in the earlier development of mathematical logic (see van Heijenoort, 1967a). These are the conceptions of *logic as universal language*, represented mainly by Frege's conceptual script, and *logic as calculus* represented mainly by the algebra of logic (see van Heijenoort, 1967a). It has been considered the standard received view of the history of symbolic logic. van Heijenoort took this distinction from Frege's own opposition between what he called lingua *characteri[sti]ca* and calculus *ratiocionator*, formulated in a paper published posthumously, in which he aimed to offer a better explanation of his intentions in writing his work *Begriffsschrift*.

With the expression *'lingua characteri[sti]ca'*, Frege meant a language with a fixed interpretation (a mathematical domain), serving "to express a content", as Frege himself wrote. His conceptual notation was conceived not only as a formal language to avoid logical errors and ambiguity, but also as a *universal scientific language*, which would "fill the gaps in the existing formula languages" and "connect their hitherto separated fields into a single domain" (Frege, 1879, pp. 7). On the contrary, a calculus was conceived only as a symbolic system without a fixed interpretation, and it was intended to be a formal representation of logic for solving logical problems. It must be taken into account that Frege wrote this comment to answer criticism dismissing his own symbolic system as a mere calculus and not as a real 'conceptual notation' or a universal language. In the opposition calculus vs. language (as least according to van Heijenoort's original ideas) a broader problem is taken into account: the problem of the role played by the *systems of signs* used in symbolic logic. In the algebraic tradition, the special signs are introduced as a *formal representation in order to solve problems*. On the contrary, for Frege the systems of signs serve as a normalization or regimentation of natural language, and they have mainly a *descriptive* function (the description of the deepest structure of thought), that takes shape in the analysis of judgments.

With his distinction, van Heijenoort wanted to underline some features that are present in Frege. Above all the *universality* of the *Begriffsschrift* should be mentioned. The logic refers to the entire universe, and there is one and fixed universe that is the universe (in words of van Heijenoort, 1967a, pp. 12). Another point stressed by van Heijenoort is the possibility of analyzing sentences, so that it can be really a *lingua*, and not like in Boole's logic, where sentences are unanalyzed and do not have a real (fixed) meaning. In the analysis Frege used his differentiation between functions expressions and object expressions as a pattern (for representation of arithmetical facts).

Van Heijenoort's distinction is *conceptual* and *normative* rather than purely historical (it would be then a philosophical distinction concerning the nature of logic). However, van Heijenoort claimed that a *dépassement* of the distinction took place after the first decades of the 20th century due to the contributions of Skolem, Herbrand and Gödel (van Heijenoort, 1967b, pp. 328). Later, the distinction was generalized by Jaakko Hintikka, who applied it on many occasions to show the different philosophical assumptions in current work in logic and to the interpretation of 20th century philosophy *tout court* (see for example the introduction of Hintikka, 1997).

Hintikka preferred to refer to the conception of language as a *universal medium* and the conception of language as *calculus*. (see Hintikka, 1997, pp. X ff. and pp. 21). He used it to describe, for example, the origins of model theory (see Hintikka, 1988). But in Hintikka the distinction reflects above all some aspects of a differentiation between theoretical and philosophical conceptions of language and logic. In this sense he used the distinction in order to make explicit what he called an "ultimate presupposition of 20th century Philosophy". This presupposition can be expressed as the question as to whether language — ordinary language — "is universal in the sense of being inescapable" (Hintikka, 1997, pp. ix). Thus, he connected the universalistic conception with idea of the ineffability of semantics, that is, that one cannot escape from the language, which is — in fact — the only and unique language possible. There is no possible metalanguage, and we can learn a language only by means of suggestions and clues. In the universalist conception, semantics cannot be defined in our only language without falling in a vicious circle. As Hintikka expresses it:

> "...the meanings of all expressions of L cannot be specified in L in one fell swoop without already presupposing the meanings of *some* expressions of L. But does not even mean that there are any particular expressions in L whose truth cannot somehow be explained by other means of expression provided by L. What it implies is that this cannot be done for all expressions of L at once." (Hintikka, 1997, pp. 35)

So, semantics cannot be expressible in the language. This fact motivated Hintikka to speak of the "ineffability of semantics". Hintikka refers to Tarski's theory of truth: In few words, the meaning of a sentence should consist in knowing the conditions to establish the sentences as true, but this knowledge should be expressed by sentences in the same language, that is, truth "is undefinable, except by transcending the language for which it has to be defined." (Hintikka, 1997, pp. 36)

3   The universalist perspective in proof-theoretic semantics

Michael Dummett provided some of the main and well known general semantic ideas which are generally accepted in proof-theoretic semantics. Summarizing, they are the following (see Dummett, 1978, pp. 222 ff.):

(i) the principle of compositionality (the meaning of a expression is determined by the meanings of its constituents),

(ii) semantic molecularism (individual sentences carry a content according to the way they are compounded), the existence of a connection between meaning and linguistic understanding.

(iii) the existence of only one key concept for the elucidation of meaning.

Dummet resorts to the notion of proof as the key concept in order to analyze the meaning of mathematical sentences (see Dummett, 1978, pp. 225 ff.). He argues his position through the following points:

(1) the meaning of an expression is determined by its use: if two expressions have the same use, then they have the same meaning. This assertion follows from the following facts: firstly, there must be observable evidence for the meaning of an expression, so that it can be communicated to others; secondly, to learn the meaning of an expression consists in learning how to use it in certain circumstances, so that our knowledge of the meaning of an expression is reduced to the knowledge of its use. Finally, the knowledge of an expression should be implicit knowledge, that is, it should be manifest in behavior. The same is valid for mathematical sentences.

(2) There must be some special aspects of the use of a sentence that constitutes its meaning. In the case of mathematics this special aspect rests on proving sentences. For in mathematics people learn what the conditions are for establishing the sentences as true.

(3) There are two aspects of the use of a sentence that are implicit in proofs: the conditions for asserting the sentences and the consequences of asserting the sentences under those conditions.

In his arguments Dummett seems to presuppose a universalist conception of language. This claim is supported by the following reasons: (a) the logical constants used by the mathematician are the logical constants of ordinary language, (b) incompleteness results due to Gödel make it impossible to really consider proofs as formulated in a formal language, (c) the use of every expression depends on observable behavior; there is no metalanguage that could describe or define it.

Some universalist features can also be found in Dag Prawitz's arguments for a proof-theoretic semantics (which in general agrees with Dummett's position). As it was remarked before, Prawitz holds a *verificationist* position about meaning and truth in general, for the *whole language* (see

Prawitz, 1998, pp. 41), and the elucidated logical constants are intended to be the logical constants of ordinary language. Furthermore, he criticizes the classical theory of truth because it makes a distinction between the semantic level of truth and the (syntactic) level of proof in a way that he calls the two-layers-view. According to it,

> "one has first to clarify the notions of meaning and truth independently of the notion of proof, and once this is done the notion of proof is easily tackled" (Prawitz, 1978, pp. 25)

Prawitz argues that truth preservation is not a sufficient condition for validity. On the contrary, the notion of meaning and proof are intertwined with each other. Hence, Prawitz proposes to *unify* the two layers in the analysis and reconstruction of proofs, the formal system being a direct representation of informal practices of proof. The analysis consists now in determining properties like harmony between introduction and elimination rules and conservativeness of extensions. To some extent, this proposal can be interpreted as the elimination of the distinction between semantic and syntactic levels, endowing formal languages with different functions: they should be something like a representation of ordinary (mathematical) language in order to prove formal results about the logic. This situation is analogous — to some extent — to the universalist views contained in Frege's *Begriffsschrift* or in *Principia Mathematica*, where the semantics is, so to say, built-in in the formal language.

4 Language from the intuitionistic perspective

As it was noted, one of the roots of proof-theoretic semantics is the BHK interpretation ("Brower-Heyting-Kolmogorov interpretation") of logical constants, dating from Heyting's reflections during the 1930s and Kolmogorov's "logic of problems" from 1932 (further details can be read in van Atten 2009). It can be argued that the interpretation was already implicit in Brouwer's ideas. Proof is the key notion to understand the meaning of logical constants. It must be noticed that the original aim of the BHK interpretation consisted only in providing an elucidation of the meaning of logical constants as they were used by the intuitionists and to make clear the differences with the classical approach. The interpretation was informal, using the constructive sense of proof in a purely informal way and no interpretation function between formalized languages was presupposed.

The original intuitionistic conception of mathematics rules out the idea of constructing formal languages. Moreover, intuitionism was very far from a linguistic conception of mathematical reasoning. According to L. E. J. Brouwer, mathematics is a mental activity, a natural function of thought, a production of the human mind. In his dissertation of 1907, mathematics was based on the subject's intuition of time and the repetition of perceived things in time (see Brouwer, 1907, pp. 81). In this framework, language played no essential role:

> "Intuitionistic mathematics is an essentially language-less mental structure" (Brouwer, 1947, pp. 339)

Mathematical knowledge is recorded in a system of signs only in order to aid memory and mathematical language is an imperfect tool for *communication*. As Heyting expressed more clearly:

> "[The intuitionist mathematician] uses language, both natural and formalized, only for communicating thoughts, i.e., to get others or himself to follow his own mathematical ideas. Such a linguistic accompaniment is not a representation of mathematics; still less is it mathematics itself." (Heyting, 1931, pp. 52 f.)

Certainly, language does not represent the mental (individual) mathematical structures, but is only a medium for *communicating* them imperfectly and in a limited form. Heyting summarizes this idea as follows:

> "We use language to influence the thoughts and actions of other people. When a mathematician writes a paper or a book, he intends to suggest mathematical constructions to other people; when he makes notes to aid his memory, his future self plays the part of another person." (Heyting, 1974, pp. 88)

In fact, Heyting was following very closely Brouwer's conception of language, and that of the *significs* movement, to which Brouwer himself belonged. This movement, quite important in the Netherlands at that time, had been initiated by Victoria Welby in England at the end of the 19th century. It was conceived as a general theory of meaning and communication related to social reform, social welfare and politics. The movement focused on the use of linguistic expressions and on the intentions of the users, borrowing ideas from the theories of signs and from different disciplines like philosophy, psychology, sociology and pedagogy. The meaning of a linguistic expression was dependent on the effect the speaker aims at or the hearer undergoes. So, rhetoric aspects and the use of language in politics were decisive issues. To this movement belonged writers and scientist from different provenance, like Henri Borel and G. Mannoury.

After World War II, Brouwer himself published an outline of the main ideas of the movement together with a description of its evolution. Originally he regarded the original task of significs as "the creation of a new stock of words bringing verbal intercourse and in consequence social organization within the reach of the spiritual tendencies of life." (Brouwer, 1946, pp. 206), but now Significs should "fight against the abuse of hysterical devices, the fight for unmasking them in private and for removing them from public life." (Brouwer, 1946, pp. 208). For example, in his opinion the state had "to use a language strictly indicative". Brouwer's ideas on Significs deserve a closer examination, but that is outside the scope of this paper. (See also Legris, 2007, further discussion on the subject can be found in chap. V of van Stigt 1990.)

Brouwer was much more concerned with the social aspects of ordinary language than its role in mathematical knowledge or its formalized versions. Moreover, his main disciple Arendt Heyting held a skeptical position concerning meaning not far from Hintikka's idea of the "ineffability" of language. In the opening dialogue of Heyting's brilliant introductory book on intuitionism, the intuitionist mathematician says:

> "If we adopt this point of view [the formalist], we clash against the obstacle of the fundamental ambiguousness of language. As the meaning of a word can never be fixed precisely enough to exclude every possibility of misunderstanding, we can never be mathematical sure that the formal system expresses correctly our mathematical thoughts." (Heyting, 1971, pp. 4)

As a consequence, the formalization of intuitionistic mathematics "will never produce a complete and definite description of it." (Heyting, 1971, pp. 9)

5   Proof-theoretic Semantics and the Role of Formalization

At this point, it seems indispensable to achieve an assessment of proof-theoretic semantics with relation to the original intutionistic ideas. It must be stressed that the general "ultimate presupposition" of 20th Century philosophy underlies proof-theoretic semantics: the validity of intuitionistic logic must be determinate in linguistic terms (in terms of the use of language). Therefore, it deviates from the original intuitionistic ideas. Furthermore, Dummett's position concerning mathematical language was inspired by Wittgenstein (who also shared the "ultimate presupposition" of

20th Century philosophy) and hence it is very different from Brouwer's view of mathematics as a "languageless" activity. So, how should the principles and results of proof-theoretic semantics be understood?

Valuable indications for a fruitful answer to this question can be found in Heyting's own assertions about the function of formalization of intuitionistic logic:

> "We may consider the formal system itself as an extremely simple mathematical structure; its entities (the signs of the system) are associated with other, often very complicated, mathematical structures. In this way formalizations may be carried out inside mathematics, and it becomes a powerful mathematical tool. Of course, one never is sure that the formal system represents fully any domain of mathematical thought: at any moment the discovery of new methods of reasoning may force us to extend the formal system." (Heyting, 1971, pp. 5)

So it cannot be ruled out that the occurrence of new methods of proof could modify the meaning given in the formal semantics to logical constants; their meaning is not fixed beforehand. Formalization would be a task accomplished *within* the existing mathematical practice:

> "From the point of view which I am sketching their importance [of the formalistic method] is mainly practical. An inconsistent system, in which every formula is derivable, cannot be very useful." (Heyting, 1974, pp. 89)

Formal language is conceived as a tool of analysis. But this sort of language is by no way "the" language; it is not everyday language nor the formalization of it. It is a system of signs that serves as a "powerful" tool for the analytical study of the intuitionistic principles and procedures, even if it cannot provide a full description of them. Thus, it should be clear that Heyting's point of view about formalization does not belong to the perspective of logic as calculus, even if it owes a lot to this perspective. In fact, Heyting formulated his ideas after the aforementioned dépassement of the distinction calculus — universal language alleged by van Heijenoort.

It is noteworthy that Heyting's position is close to another position in formal sciences: the tradition of *symbolic knowledge* steaming mainly from G. W. Leibniz. In few words, symbolic knowledge is knowledge obtained by means of a semiotic structure, where there is a procedure for the production of complex signs on the basis of basic signs according to a set of rules in a *very* general sense. Besides, no presupposition is made about the nature of these signs or what they represent. The idea of symbolic knowledge is *pragmatic* in essence, and it does not involve only a semantic purpose (see Esquisabel, 2012). The *locus classicus* where the expression 'symbolic knowledge' appears is Leibniz's "Meditationes de cogitationes, veritate et ideis" from 1684 (see A VI 4, 587-588; GP IV, 423). Through this notion, Leibniz aimed at a justification of the epistemological use of calculi and artificial languages, that is, the use of calculi and artificial languages as a tool to gain new knowledge.

With the introduction of this notion of symbolic knowledge an important methodological innovation was achieved: the knowledge obtained through *symbolic manipulation*, produced in the form of calculi. Notwithstanding, symbolic knowledge includes also a *representational function* of symbolic systems. Representation must be understood here in the sense of mappings or morphism from a structure to another, allowing what is called *surrogative* reasoning. This function of exhibition by symbols was called by Leibniz *ecthetic*. So, there are two aspects of symbolic knowledge: the *computational* and the *structural* (for further details, see Legris, 2012). The notion of symbolic knowledge initiated a "tradition of symbolic knowledge" in the methodology of formal sciences. After Leibniz other philosophers and scientists of the 18th and 19th centuries developed this notion in different directions.

Symbolic knowledge was presupposed in the contributions of 19th century algebra of logic (see Legris, 2012, pp. 84 ff.), and from then on it pervaded the further evolution of mathematical logic and the formal approach in foundations of mathematics. The two mentioned aspects of symbolic knowledge are implicit in Heyting's quotations. Hence, his ideas can be interpreted as the adoption of the methodology of symbolic knowledge for the examination of intuitionistic logic and mathematics. When paving the way for the formal analysis of intuitionistic logic, Heyting's ideas made proof-theoretic semantics possible. Taking this into account, proof-theoretic semantics should not be regarded as a natural and direct outcome of intuitionism, but as a way to show the mathematical structure underlying the formalization of intuitionistic logic. Of course, this claim can encourage further discussion.

References

Brouwer, L. E. J. (1907), Over de grondslagen der wiskunde, PhD thesis, Amsterdam. Reprinted Amsterdam, Mathematisch Centrum, 1981. Engl. Transl. in Collected Works Vol. I, ed. by Arendt Heyting. Amsterdam, North-Holland, 1975, pp. 11–101.

Brouwer, L. E. J. (1946), 'Synopsis of the signific movement in the nether-lands. prospects of the signific movement', *Synthese* 5, 201–208.

Brouwer, L. E. J. (1947), Richtlinien der intuitionistische wiskunde, *in* 'Proc. Akad. Amsterdam', Vol. 50, p. 339.

Dummett, M. (1978), *Truth on Other Enigmas*, Duckworth, chapter The Philosophical Basis of Intuitionistic Logic, pp. 215–247.

Esquisabel, O. M. (2012), *Lassalle Casanave*, chapter Representing and abstracting. An Analysis of Leibniz's Concept of Symbolic Knowledge, pp. 1–49.

Frege, G. (1879), *Begriffsschrift*, Louis Nebert. Reprint in Ignacio An-gelelli (ed). Gottlob Frege. Begriffsschrift und andere Aufsätze. Mit E. Husserls und H. Scholz' Anmerkungen. Darmstadt, Wissenschaftliche Buchgesellschaft,1964. Engl. transl. in van Heijenoort (1967a).

Gentzen, G. (1934), 'Untersuchungen über das logische schließen', *Mathematische Zeitschrift* 39, 176–210 and 405–431. Eng. Transl.: "Investigations into Logical Deduction", In Collected Papers ed. and transl. By M. E. Szabo. Amsterdam – London, North-Holland, 1969, pp. 68–131.

Heyting, A. (1931), 'Die intuitionistische grundlegung der mathematik', *Erkenntnis* 2, 106–115. Engl. Translation "The Intuitionistic Foundations of Mathematics". In Benacerraf, Paul & Putnam, Hillary (eds.) Philosophy of Mathematics. Selected Readings. 2nd. Ed. Cambridge, Cambridge University Press, 1983, pp. 53–61.

Heyting, A. (1971), *Intuitionism. An Introduction*, North-Holland.

Heyting, A. (1974), 'Intuitionistic views on the nature of mathematics', *Synthese* 27, 79–91.

Hintikka, J. (1988), 'On the development of the model-theoretical viewpoint in logical theory', *Synthese* 77, 1–36.

Hintikka, J. (1997), *Lingua Universalis vs. Calculus Ratiocinator. An Ultimate Presupposition of Twentieth-Century Philosophy*, Dordrecht et al., Kluwer.

Legris, J. (2007), Cálculo y lenguaje en el intuicionismo matemático, *in* 'Anales de la Academia Nacional de Ciencias de Buenos Aires', Vol. XLI, pp. 943–952.

Legris, J. (2012), Between calculus and semantic analysis, *in* L. Casanave, ed., 'Symbolic Knowledge in the Origins of Mathematical Logic', pp. 79–113.

Prawitz, D. (1978), *Essays on Mathematical and Philosophical Logic*, Reidel, chapter Proofs and the Meaning and Completeness of the Logical Constants, pp. 25–40.

Prawitz, D. (1998), *Truth in Mathematics*, Clarendon Press, chapter Truth and Objectivity from a Verificacionist Point of View, pp. 41–51.

Prawitz, D. (2006), 'Meaning approached via proofs', *Synthese* **148**, 507–524.

Schroeder-Heister, P. (2012), *The Stanford Encyclopedia of Philosophy (Summer 2014 Edition)*, chapter Proof-Theoretic Semantics. URL http://plato.stanford.edu/archives/sum2014/entries/proof-theoretic-semantics/.

van Heijenoort, J. (1967a), *From Frege to Gödel. A Source Book in Mathematical Logic*, Harvard University Press.

van Heijenoort, J. (1967b), 'Logic as calculus and logic as language', *Synthese* **24**, 324–330. Reprinted in van Heijenoort 1985, pp. 11–16.

On the ordered Dedekind real numbers in toposes[‡§]

Marcelo E. Coniglio*   Luís A. Sbardellini*

* Centre for Logic, Epistemology and the History of Science (CLE)
State University of Campinas (UNICAMP), Campinas, Brazil
coniglio@cle.unicamp.br
sbardellini@gmail.com

Abstract

In 1996, W. Veldman and F. Waaldijk presents a constructive (intuitionistic) proof for the homogeneity of the ordered structure of the Cauchy real numbers, and so this result holds in any topos with natural number object. However, it is well known that the real numbers objects obtained by the traditional constructions of Cauchy sequences and Dedekind cuts are not necessarily isomorphic in an arbitrary topos with natural numbers object. Consequently, Veldman and Waaldijk's result does not apply to the ordered structure of Dedekind real numbers in toposes. The main result to be proved in the present paper is that the ordered structure of the Dedekind real numbers object is homogeneous, in any topos with natural numbers object. This result is obtained within the framework of local set theory.

1   Introduction: From Intuitionism to Toposes

As it is well-known, Intuitionism was introduced by L.E.J. Brouwer in 1907 (Brouwer, 1907) as a philosophy of mathematics based on the idea that mathematics is a creation of the mind. From this perspective, the truth of mathematical statements is stated by means of *mental constructions*. In particular, the principle of the excluded middle ($A \vee \neg A$) is rejected under the intuitionistic point of view because of the lack of *constructivistic* character.

The main logical principles of Intuitionism were formalized by A. Heyting through the so-called Intuitionistic logic, in which, by the reasons

---

‡Dedicated to Luiz Carlos Pereira on the occasion of his 60th birthday

§We would like to thank Wim Veldman and Ricardo Bianconi for several helpful suggestions and remarks made in an early stage of this project. This research was supported by FAPESP (Brazil) and CNPq (Brazil).

pointed out above, the third-excluded law is no longer valid. Intuitionistic logic was adopted as the logical framework underlying the *constructive mathematics*, including the constructive analysis of E. Bishop and the recursive analysis developed in Russia.

Some decades after the introduction of Intuitionism, S. Eilenberg and S. MacLane proposed in 1945 the basis of a structuralistic approach to mathematics via the theory of *categories and functors* (Eilenberg & MacLane, 1945). The main feature of category theory is that the concepts of structure ('object') and morphism ('arrow') are taken as primitive.

Category theory meets Intuitionism through the notion of *topos*. Toposes are particular categories endowed with enough structure which allows to consider them as a kind of *mathematical universes* or *universes of sets*. They constitute models of a higher-order intuitionistic type theory, where the subobject classifier Ω corresponds to a type of (intuitionistic) truth values. Being so, it is usually claimed that the internal logic of toposes is intuitionistic. This is rigorously supported by the Mitchell-Bénabou language defined on toposes, together with the associated Kripke-Joyal semantics (see, for instance, MacLane & Moerdijk (1992), Chapter VI).

In 1988, J.L. Bell introduced *Local Set Theory* (see Bell (1988)) as another formal (or logical) counterpart of the notion of toposes, formulated in a typed intuitionistic logic. It can be considered as a generalization of classical set theory such that the category of sets can always be obtained, and shown to be a topos (as in the 'classical', set-theoretic framework). Moreover, any topos can be obtained as the category of sets within a suitable local set theory, which exposes to what extent toposes can be considered as a generalization of the categories of sets.

The development of (set-theoretic) Model Theory for Intuitionistic logic was already considered in the literature (see, for instance, Fitting (1969), de Swart (1978), Veldman & Waaldijk (1996) and more recently Constable & Bickford (2014)). Given the close relationship between toposes and intuitionistic logic pointed out above, it seems natural to analyze intuitionistic model theory by using the framework of local set theory (see Sbardellini & Coniglio (2006) and Sbardellini (2005)).

This paper, continuing our previous research, deals with the real numbers object obtained in a topos with natural number object, constructed by means of a generalization of the Dedekind techniques. The intuitive meaning of the Dedekind real numbers defined, in particular, in the emblematic topos $\mathbf{Sh}(X)$ of sheaves over a topological space X, compared with the usual construction in the topos **Set** of sets, was clearly described by Bell:

Consider, for example, the concept of 'real-valued continuous func-

tion on a topological space X'. Any such function may be regarded as a real number (or quantity) varying continuously over X. Now consider the topos $\mathbf{Sh}(X)$ of sheaves on X. Here everything is varying (continuously) over X, so shifting from $\mathbf{Set}$ to $\mathbf{Sh}(X)$ essentially amounts to placing oneself in a framework which is, so to speak, itself 'co-moving' with the variation over X of a given variable real number. This causes its variation not to be 'noticed' in $\mathbf{Sh}(X)$; it is accordingly regarded as being a constant real number. In this way the concept of 'real-valued continuous function on X' is transformed into the concept of 'real number' when interpreted in $\mathbf{Sh}(X)$. [...] Putting in the other way round, the concept 'real number', interpreted in $\mathbf{Sh}(X)$ corresponds to the concept 'real-valued continuous function on X' interpreted in $\mathbf{Set}$.

J. L. Bell, Bell (1988), pp. 240.

The idea of quantities varying continuously can be connected to the model-theoretic concept of *homogeneity*. The classical model-theoretic notion of homogeneity of a structure was introduced by B. Jónsson in 1960 (Jónsson, 1960). It can be formulated as follows: If κ is an infinite cardinal, a set-based structure $\mathfrak{A}$ is said to be κ-homogeneous if, for every partial isomorphism of $\mathfrak{A}$ of cardinal $\kappa' < \kappa$, there exists an automorphism of $\mathfrak{A}$ which extends it (see Bell & Slomson (1971) for details). In this paper we only deal with finite partial isomorphisms, which means that our homogeneous ordered structures are the $\aleph_0$-homogeneous ones.

Within a set-theoretic approach, Veldman & Waaldijk (1996) presents a constructive (intuitionistic) proof for the homogeneity of the ordered structure of the Cauchy real numbers, and so this result holds in any topos with natural number object. It is well known that, in an arbitrary topos with natural numbers object, the real numbers objects obtained by the traditional constructions of Cauchy sequences and Dedekind cuts are not necessarily isomorphic (see, for instance, Mulvey (1974) and Johnstone (1977)), and so as for the corresponding ordered structures. Consequently, the result obtained in Veldman & Waaldijk (1996) does not apply to the ordered structure of Dedekind real numbers in toposes. The main result to be proved in the present paper is that the ordered structure $\langle \mathbb{R}_d, < \rangle$ of the Dedekind real numbers object is homogeneous, in any topos with natural numbers object (cf. Theorem 5.2 below).

In Sbardellini & Coniglio (2006) we introduced the concept of effectively homogeneous ordered structures, for which there is an effective procedure which extends every finite partial isomorphism to an automorphism. The present paper follows closely the formal treatment developed in Sbardellini & Coniglio (2006), based on the logical framework of local set theory (cf. Bell (1988, n.d.)). As a consequence of this constructive approach, it will be shown that the structure $\langle \mathbb{R}_d, < \rangle$ is effectively homogeneous. The full details of this constructions can be found in Sbardellini (2005).

Recall that local set theory is a typed set theory, presented through a sequent calculus S, whose underlying logic is (many-sorted) higher-order intuitionistic logic; in this manner, the primitive notion of set is replaced by that of terms of *power* types. The resulting local sets (or S-sets) and arrows (or S-maps) set up a category $\mathbf{C}(S)$ that can be shown to be a topos, called a *linguistic topos*. It can to show that every topos $\mathbf{E}$ is equivalent to a linguistic topos, namely $\mathbf{C}(\mathrm{T}(\mathbf{E}))$, where $\mathrm{T}(\mathbf{E})$ is the local set theory whose axioms are those which are valid in the canonical interpretation of the internal language of $\mathbf{E}$ into $\mathbf{E}$ itself. Thus, the categorial machinery of a topos can be translated into a logical one and so we may develop all the technical constructions within the environment of local set theory.

The layout of the paper is the following: In Section 2 we summarize some basic definitions and results previously introduced in Sbardellini & Coniglio (2006), mostly concerned with local set theory and ordered structures; in Section 3 we define Dedekind cuts and establish the necessary background for the further proof of homogeneity; in Section 4 we briefly recall the main results concerning natural numbers and finite sequences which was exhibited in Sbardellini & Coniglio (2006); finally, it is proved in Section 5 the main result of this paper (Theorem 5.2): the ordered structure of the Dedekind real numbers is (effectively) homogeneous in any topos with natural numbers object.

2 Preliminaries on Local Set Theories

As mentioned in the previous section, the framework we adopt here is that of local set theories. For that reason, these preliminaries will serve specifically as a notational guide for the other sections. We address the reader to Bell (1988) (see also Bell (n.d.)) for a more detailed exposition of the subject of local set theories.

We begin by a *local language* $\mathcal{L}$, which is a higher-order language consisting of types, variables and terms defined as usual, with the following relevant features: if $A_1, \ldots, A_n$ and A are types then $A_1 \times \cdots \times A_n$ and PA are types (the *product* and the *power* types, respectively). If $n = 1$ then $A_1 \times \cdots \times A_n$ is A_1. On the other hand, if $n = 0$ then the empty product $A_1 \times \cdots \times A_n$ is denoted by $\mathbf{1}$ (the *unity* type). There is just one more distinguished type, the *truth-value* type, denoted by Ω. For every type A we have a denumerable set $x_A^1, x_A^2, \ldots$ of *variables* of type A. The set of *terms* of a given type A is defined recursively over the set of variables. The details of the construction can be found in the book Bell (1988) (see also Bell (n.d.)). The terms of type Ω are called *formulas* and denoted by α, β etc. A formula *in context* is an expression $\vec{x}.\alpha$, where $\vec{x}$ is a list $x_0, \ldots, x_{n-1}$ of distinct variables and α is a formula such that all its free variables are in $\vec{x}$.

Local sets are defined as being the closed terms of power type. They will be denoted by capital letters A, B, X, Y etc. Recall from Bell (1988, n.d.) that, if x is a variable of type A and t is a term of type PA then $(x \in t)$ is a formula.[1] If α is a formula then $\{x : \alpha\}$ is a term of type PA in which every occurrence of the variable x (of type A) is not free. This means that, in a local language, there is a binding operator $\{ : \}$ for every type A. As in Sbardellini & Coniglio (2006), we will represent elements of a local set with the same letter of the local set to which they belong, though in low case letters (possibly indexed). For example: $a, a', a_0 \in A$; $b', b'', b_1 \in B$; $x, x_1, x_2 \in X$ and so on. In a local language $\mathcal{L}$, all the customary logical symbols (including connectives and other binding operators, such as quantifiers) can be *defined* using the primitive symbols of the language. Two special local sets deserve a mention: $U_A = \{x_A : \top\}$ and $\emptyset_A = \{x_A : \bot\}$, for every type A.

A *local set theory* is a sequent calculus S over a local language $\mathcal{L}$ satisfying specific rules for higher-order intuitionistic logic.[2] The notation and terminology we will use in the sequel is standard and does not differ essentially from Bell (1988). For instance, we will write $\Gamma \vdash_S \alpha$ to denote that the sequent $\Gamma \Rightarrow \alpha$ is derived from the collection of sequents S. Now we recall some elementary facts about local set theories. An S-set X is said to be *inhabited* if $\vdash_S \text{INH}(X)$, where $\text{INH}(X) :\Leftrightarrow \exists x.x \in X$. We also say that a formula in context $\vec{z}.\alpha$ is *decidable in* the S-set $\prod_{i<n} X_i$, where $X_i \subseteq U_{A_i}$ (for $i = 0, \ldots, n-1$), if

$$\vdash_S \forall x_0 \in X_0, \ldots, x_{n-1} \in X_{n-1}[\alpha \vee \neg\alpha].$$

Equivalently, $\vec{z}.\alpha$ is decidable in $\prod_{i<n} X_i$ if the S-set

$$\left\{ \langle x_0, \ldots, x_{n-1} \rangle \in \prod_{i<n} X_i : \alpha \right\}$$

is a complemented element in the Heyting algebra of the S-subsets of $\prod_{i<n} U_{A_i}$. In particular, $z.\alpha$ is decidable in $X \subseteq U_A$ if $\vdash_S \forall x \in X[\alpha \vee \neg\alpha]$.

---

[1]To be strict, there is a symbol $\in_A$ for every type A.

[2]We assume here that the reader is acquainted with the basic axioms and inference rules of higher-order intuitionistic logic.

Moreover, we say that $\vec{z}.\alpha$ is *decidable* if it is decidable in $\prod_{i<n} U_{A_i}$. If $\vec{z}.\alpha$ and $\vec{z}.\beta$ are decidable in $\prod_{i<n} X_i$, then so are $\vec{z}.(\alpha \wedge \beta)$, $\vec{z}.(\alpha \vee \beta)$ and $\vec{z}.(\alpha \to \beta)$ (see Sbardellini & Coniglio (2006)).

Next we enunciate some basic definitions concerning ordered structures in a local set theory, taken from Sbardellini & Coniglio (2006). Recall that $\mathbf{C}(S)$ is the topos constructed from the local set theory S. A (*partially*) *ordered* $\mathbf{C}(S)$-*structure* (in short, "an order") is a pair $\langle A, < \rangle$, where A is an S-set and $< \subseteq A \times A$ is a relation satisfying

$$\vdash_S \neg(a < a) \quad \text{and} \quad (a' < a''), (a'' < a''') \vdash_S (a' < a''').$$

A *homomorphism* f from $\langle A, < \rangle$ into $\langle B, < \rangle$ is defined to be an S-map $f : A \to B$ which preserves the order. Ordered $\mathbf{C}(S)$-structures and homomorphisms form a category denoted by $\mathrm{Ord}[\mathbf{C}(S)]$. We say that an order $\langle A, < \rangle$ is *linear* if it satisfies

$$\vdash_S (a' < a'') \vee (a' = a'') \vee (a'' < a').$$

An order $\langle A, < \rangle$ is said to be *dense in* $\langle B, < \rangle$ if there exists a monomorphism $i : \langle A, < \rangle \to \langle B, < \rangle$ in $\mathrm{Ord}[\mathbf{C}(S)]$ such that

$$b' < b'' \vdash_S \exists a.b' < i(a) < b''.$$

The monomorphism i will be clear from the context. For example, in **Set** this monomorphism is represented in most cases by the inclusion $A \subseteq B$. An order $\langle A, < \rangle$ is *dense* if it is dense in itself by means of id_A.

An order $\langle A, < \rangle$ is *persistent in* an order $\langle B, < \rangle$ if it is dense in $\langle B, < \rangle$ and

$$\vdash_S \forall b \exists a', a''.i(a') < b < i(a''),$$

where i is the monomorphism mentioned in the definition of density. Finally, an order $\langle A, < \rangle$ is *persistent* if it is persistent in itself. Intuitively, $\langle A, < \rangle$ is persistent if it is dense and does not have endpoints.

3　Dedekind cuts in Local Set Theories

In classical set theory, Dedekind cuts are presented as constructions performed exclusively over the set of rational numbers. This approach is supported by Cantor's back and forth theorem, which also holds in any topos with natural numbers object (cf. Sbardellini & Coniglio (2006)). Nevertheless, the construction can be generalized to arbitrary linearly ordered structures in order to embrace pure toposes, not necessarily containing a natural numbers object.

As in the classical case, a *Dedekind cut over* a linearly ordered $C(S)$-structure $\mathfrak{A} = \langle A, < \rangle$ is a pair $\langle X, Y \rangle \in PA \times PA$ satisfying

$$\mathrel{\vdash\kern-0.6em\sim}_S \mathrm{DED}_{\mathfrak{A}}(\langle X, Y \rangle),$$

where

$$
\begin{aligned}
\mathrm{DED}_{\mathfrak{A}}(\langle X, Y \rangle) :\Leftrightarrow\ & \mathrm{INH}(X) \wedge \mathrm{INH}(Y) \wedge X \cap Y = \emptyset \\
& \wedge\ \forall a'(a' \in X \leftrightarrow \exists a'' \in X.a' < a'') \\
& \wedge\ \forall a'(a' \in Y \leftrightarrow \exists a'' \in Y.a'' < a') \\
& \wedge\ \forall a', a''(a' < a'' \to a' \in X \vee a'' \in Y).
\end{aligned}
\tag{1}
$$

We represent the S-set of all the Dedekind cuts of $\mathfrak{A}$ by ∂A. Thus:

$$\partial A := \{ u \in PA \times PA : \mathrm{DED}_{\mathfrak{A}}(u) \}.$$

By defining a strict order over ∂A by

$$u < v :\Leftrightarrow u, v \in \partial A \wedge \exists a.a \in \pi'(v) \cap \pi''(u),
\tag{2}$$

we obtain a $C(S)$-structure $\langle \partial A, < \rangle$, which will be denoted by $\partial \mathfrak{A}$. (In the expression above, π' and π'' denote the canonical projections.) The next lemma assures that the construction by Dedekind cuts preserves isomorphisms between $C(S)$-structures.

Lemma 3.1. *If $\mathfrak{A} = \langle A, < \rangle$ and $\mathfrak{A} = \langle B, < \rangle$ are isomorphic linearly ordered $C(S)$-structures, then $\partial \mathfrak{A}$ and $\partial \mathfrak{B}$ are isomorphic.*

Proof. Let $h : \mathfrak{A} \to \mathfrak{B}$ be such an isomorphism. First observe that

$$\mathrel{\vdash\kern-0.6em\sim}_S \forall u \in \partial A.\mathrm{DED}_{\mathfrak{B}}(\langle h \circ \pi'(u), h \circ \pi''(u) \rangle).$$

Now define $\tilde{h} : \partial A \to \partial B$ by

$$u \mapsto \langle h \circ \pi'(u), h \circ \pi''(u) \rangle.$$

Since h is bijective, we have

$$\mathrel{\vdash\kern-0.6em\sim}_S \forall V \in PB\, \exists! U \in PA.h(U) = V$$

and consequently

$$\mathrel{\vdash\kern-0.6em\sim}_S \forall v \in \partial B\, \exists! u \in \partial A[h \circ \pi'(u) = \pi'(v) \wedge h \circ \pi''(u) = \pi''(v)],$$

which shows that $\tilde{h}$ is a bijection. It remains to show that $\tilde{h}$ preserves the relation $<$. In fact:

$$
\begin{aligned}
u, v \in \partial A \mathrel{\vdash_S} u < v \;&\leftrightarrow\; \exists a. a \in \pi'(v) \cap \pi''(u) \\
&\leftrightarrow\; \exists a. h(a) \in h \circ \pi'(v) \cap h \circ \pi''(u) \\
&\leftrightarrow\; \exists b. b \in h \circ \pi'(v) \cap h \circ \pi''(u) \\
&\leftrightarrow\; \tilde{h}(u) < \tilde{h}(v),
\end{aligned}
$$

where the first and second lines follow from the definition of $<$, see (2) above. $\qquad\square$

The following lemma guarantees that, if the $\mathbf{C}(S)$-structure $\mathfrak{A}$ is persistent, then there exists a monomorphism $i_{\mathfrak{A}} : \mathfrak{A} \hookrightarrow \partial \mathfrak{A}$.

Lemma 3.2. *Let $\mathfrak{A} = \langle A, < \rangle$ be a linearly ordered, persistent and inhabited $\mathbf{C}(S)$-structure. Then*

$$
\vdash_S \forall a. \mathrm{DED}_{\mathfrak{A}}(\langle \{a' \; : \; a' < a\}, \{a' \; : \; a < a'\}\rangle)
$$

and, moreover, the morphism $i_{\mathfrak{A}} : \mathfrak{A} \to \partial \mathfrak{A}$, defined by

$$
a \mapsto \langle \{a' \; : \; a' < a\}, \{a' \; : \; a < a'\}\rangle,
$$

is a monomorphism.

Proof. The first part follows directly from the definition of Dedekind cut, recall (1) above: both sets are inhabited (because $\mathfrak{A}$ does not have endpoints), disjoint, open (because $\mathfrak{A}$ is dense) and consecutive (because $\mathfrak{A}$ is linear). The second part follows from

$$
\begin{aligned}
i_{\mathfrak{A}}(a_0) = i_{\mathfrak{A}}(a_1) \mathrel{\vdash_S} \; &\langle \{a' \; : \; a' < a_0\}, \{a' \; : \; a_0 < a'\}\rangle \\
&= \langle \{a' \; : \; a' < a_1\}, \{a' \; : \; a_1 < a'\}\rangle \\
&[a' < a_0 \leftrightarrow a' < a_1] \wedge [a_0 < a' \leftrightarrow a_1 < a'] \\
&a_0 = a_1,
\end{aligned}
$$

where the last line is a consequence of the linearity of $\mathfrak{A}$. $\qquad\square$

We call $i_{\mathfrak{A}}$ the canonical monomorphism associated to $\mathfrak{A}$. This notation will be kept from now on, with possible omission of the index $\mathfrak{A}$.

Lemma 3.3. *Let $\mathfrak{A} = \langle A, < \rangle$ and $\mathfrak{B} = \langle B, < \rangle$ be two linearly ordered and persistent $\mathbf{C}(S)$-structures. Then, for every isomorphism $h : \mathfrak{A} \to \mathfrak{B}$, there exists an isomorphism $\tilde{h} : \partial \mathfrak{A} \to \partial \mathfrak{B}$ which extends h, that is, for which $\tilde{h} \circ i_{\mathfrak{A}} = i_{\mathfrak{B}} \circ h$.*

Proof. Repeating the steps of the proof of Lemma 3.1, we obtain from the isomorphism $h : \mathfrak{A} \to \mathfrak{B}$ an isomorphism $\tilde{h} : \partial\mathfrak{A} \to \partial\mathfrak{B}$ defined by

$$u \mapsto \langle h \circ \pi'(u), h \circ \pi''(u) \rangle.$$

It remains to show that $\tilde{h}$ extends h:

$$
\begin{aligned}
\vdash_S \tilde{h} \circ i_{\mathfrak{A}}(a) &= \langle h(\{a' \; : \; a' < a\}), h(\{a' \; : \; a < a'\}) \rangle \\
&= \langle \{h(a') \; : \; a' < a\}, \{h(a') \; : \; a < a'\} \rangle \\
&= \langle \{h(a') \; : \; h(a') < h(a)\}, \{h(a') \; : \; h(a) < h(a')\} \rangle \\
&= \langle \{b \; : \; b < h(a)\}, \{b \; : \; h(a) < b\} \rangle \\
&= i_{\mathfrak{B}} \circ h(a).
\end{aligned}
$$

$\square$

Concerning uniform persistence, we obtain the following result.

Proposition 3.4. *If the linearly ordered $\mathbf{C}(S)$-structure $\mathfrak{A} = \langle A, < \rangle$ is persistent, then it is uniformly persistent in $\partial\mathfrak{A}$.*

Proof. First we must show that $\mathfrak{A}$ is persistent in $\partial\mathfrak{A}$. The following expression reveals that the former is dense in the latter (recall that i is the canonical monomorphism from Lemma 3.2):

$$
\begin{aligned}
u, v \in \partial A \vdash_S u < v \;\; &\to \exists a.a \in \pi'(v) \cap \pi''(u) \\
&\to \exists a, a', a''[a' < a < a'' \;\; \wedge \;\; a', a'' \in \pi'(v) \cap \pi''(u)] \\
&\to \exists a.u < i(a) < v,
\end{aligned}
$$

where the first line follows from the definition of $<$ (see (2) above), the second from the openness of the cut, and the third again from the definition (2) of $<$. The persistence of $\mathfrak{A}$ in $\partial\mathfrak{A}$ is proved by:

$$
\begin{aligned}
u \in \partial A \vdash_S \;\; &\exists a', a''[a' \in \pi'(u) \;\; \wedge \;\; a'' \in \pi''(u)] \\
\vdash_S \;\; &\exists a', a''.i(a') < u < i(a''),
\end{aligned}
$$

in which we used the fact that "cuts are inhabited" (first line) and the definition (2) of $<$ (second line). Finally, since $\mathfrak{A}$ is persistent, we have as a particular case of Lemma 3.3 that every automorphism from $\mathfrak{A}$ into itself can be extended to an automorphism from $\partial\mathfrak{A}$ into itself. Hence we conclude that $\mathfrak{A}$ is uniformly persistent in $\partial\mathfrak{A}$. $\square$

The proposition bellow prescribes a sufficient condition for a given substructure to reach its base-structure with respect to cuts.

Proposition 3.5. *If $\mathfrak{A}$ and $\mathfrak{B}$ are linearly ordered $\mathbf{C}(S)$-structures such that $\mathfrak{B}$ is dense in $\mathfrak{A}$, then $\partial\mathfrak{B}$ is isomorphic to $\partial\mathfrak{A}$.*

Proof. Let $i : \mathfrak{B} \hookrightarrow \mathfrak{A}$ be a monomorphism in $\mathrm{Ord}[\mathrm{C}(S)]$ such that

$$a' < a'' \mathbin{\vdash_S} \exists b. a' < i(b) < a''.$$

We want to construct an isomorphism $h : \partial\mathfrak{A} \to \partial\mathfrak{B}$ from i. Note that

$$u \in \partial A \mathbin{\vdash_S} \langle \{b \ : \ i(b) \in \pi'(u)\}, \{b \ : \ i(b) \in \pi''(u)\} \rangle \in \partial B$$

because $\mathfrak{B}$ is dense in $\mathfrak{A}$. The S-function $h : \partial A \to \partial B$, defined by

$$u \mapsto \langle \{b \ : \ i(b) \in \pi'(u)\}, \{b \ : \ i(b) \in \pi''(u)\} \rangle,$$

becomes a natural choice. Now we verify that h is in fact a bijection. First we show that it is an injection. Note that

$$u, v \in \partial A \mathbin{\vdash_S} h(u) = h(v) \ \to \{b \ : \ i(b) \in \pi'(u)\} = \{b \ : \ i(b) \in \pi'(v)\}$$
$$\to \forall b[i(b) \in \pi'(u) \leftrightarrow i(b) \in \pi'(v)].$$

On the other hand,

$$\forall b[i(b) \in \pi'(u) \leftrightarrow i(b) \in \pi'(v)] \mathbin{\vdash_S} \ a \in \pi'(u)$$
$$\to \exists a'[a < a' \wedge a' \in \pi'(u)]$$
$$\to \exists b'[a < i(b') < a' \wedge i(b') \in \pi'(u)]$$
$$\to \exists b'[a < i(b') \wedge i(b') \in \pi'(v)]$$
$$\to a \in \pi'(v).$$

Similarly:

$$\forall b[i(b) \in \pi'(u) \leftrightarrow i(b) \in \pi'(v)] \mathbin{\vdash_S} a \in \pi'(v) \to a \in \pi'(u),$$

thus:

$$u, v \in \partial A \mathbin{\vdash_S} h(u) = h(v) \ \to \forall a[a \in \pi'(u) \leftrightarrow a \in \pi'(v)]$$
$$\to \pi'(u) = \pi'(v).$$

Repeating the argument for π'' we obtain:

$$u, v \in \partial A \mathbin{\vdash_S} h(u) = h(v) \to \pi''(u) = \pi''(v).$$

Finally:

$$u, v \in \partial A \mathbin{\vdash_S} h(u) = h(v) \to u = v,$$

showing that h is an injection. Now, to prove that h is surjective, we observe that:

$$\mathbin{\vdash_S} \ \pi' \circ h(\langle \{a \ : \ \exists b \in \pi'(v).a < i(b)\}, \{a \ : \ \exists b \in \pi''(v).i(b) < a\} \rangle)$$
$$= \{b \ : \ i(b) \in \{a \ : \ \exists b \in \pi'(v).a < i(b)\}\}$$
$$= \{b \ : \ \exists b' \in \pi'(v).i(b) < i(b')\}$$
$$= \{b \ : \ \exists b' \in \pi'(v).b < b'\}$$
$$= \pi'(v).$$

Similarly:

$$\mathrel{\vdash_S} \pi'' \circ h((\{a \;:\; \exists b \in \pi'(v).a < i(b)\}, \{a \;:\; \exists b \in \pi''(v).i(b) < a\})) = \pi''(v).$$

Thus:

$$\mathrel{\vdash_S} h((\{a \;:\; \exists b \in \pi'(v).a < i(b)\}, \{a \;:\; \exists b \in \pi''(v).i(b) < a\})) = v.$$

Therefore we conclude that:

$$v \in \partial B \mathrel{\vdash_S} \exists u \in \partial A.h(u) = v.$$

It remains to show that h is an isomorphism. On the one hand:

$$
\begin{aligned}
u, v \in \partial A \mathrel{\vdash_S} u < v \;
&\to \exists a.a \in \pi'(v) \cap \pi''(u) \\
&\to \exists a, a'[a < a' \wedge a, a' \in \pi'(v) \cap \pi''(u)] \\
&\to \exists b.i(b) \in \pi'(v) \cap \pi''(u) \\
&\to \exists b.b \in \{b' \;:\; i(b') \in \pi'(v)\} \\
&\qquad\qquad \cap \{b' \;:\; i(b') \in \pi''(u)\} \\
&\to \exists b.b \in \pi' \circ h(v) \cap \pi'' \circ h(u) \\
&\to h(u) < h(v)
\end{aligned}
$$

and, on the other hand:

$$
\begin{aligned}
u, v \in \partial A \mathrel{\vdash_S} h(u) < h(v) \;
&\to \exists b.b \in \pi' \circ h(v) \cap \pi'' \circ h(u) \\
&\to \exists b.b \in \{b' \;:\; i(b') \in \pi'(v)\} \\
&\qquad\qquad \cap \{b' \;:\; i(b') \in \pi''(u)\} \\
&\to \exists b.i(b) \in \pi'(v) \cap \pi''(u) \\
&\to \exists a.a \in \pi'(v) \cap \pi''(u) \\
&\to u < v.
\end{aligned}
$$

$\square$

An interesting consequence from Propositions 3.4 and 3.5 is that, if $\mathfrak{A}$ is a linearly ordered and persistent $\mathbf{C}(S)$-structure and $\partial\mathfrak{A}$ is also linear, then $\partial\partial\mathfrak{A}$ is isomorphic to $\partial\mathfrak{A}$. This means that the Dedekind cuts construction is idempotent, in some sense, when applied to persistent $\mathbf{C}(S)$-structures.

The next lemma sums up most of precedent results.

Lemma 3.6. *Consider the following conditions: (i) $\mathfrak{A}$ and $\mathfrak{B}$ are linearly ordered and persistent $\mathbf{C}(S)$-structures; (ii) f is a partial isomorphism from $\mathfrak{A}$ into $\mathfrak{B}$; (iii) $\mathrm{dom}(f)$ is dense in $\mathfrak{A}$ and $\mathrm{cod}(f)$ is dense in $\mathfrak{B}$. Then there exists an isomorphism $h : \partial\mathfrak{A} \to \partial\mathfrak{B}$ which extends f, that is, for which $h \circ i_{\mathfrak{A}}|_{\mathrm{dom}(f)} = i_{\mathfrak{B}} \circ f$.*

Proof. Let $\mathfrak{A}' = \langle \mathrm{dom}(f), <_{\mathfrak{A}} \rangle$ and $\mathfrak{B}' = \langle \mathrm{cod}(f), <_{\mathfrak{B}} \rangle$ be the respective (isomorphic) $C(S)$-structures generated by $\mathrm{dom}(f)$ and $\mathrm{cod}(f)$; we infer easily from the hypothesis that $\mathfrak{A}'$ and $\mathfrak{B}'$ are persistent. From Lemma 3.3, there exists an isomorphism $\tilde{f} : \partial\mathfrak{A}' \to \partial\mathfrak{B}'$ which extends f. Now, from Proposition 3.5, there exist isomorphisms $h' : \partial\mathfrak{A} \to \partial\mathfrak{A}'$ and $h'' : \partial\mathfrak{B} \to \partial\mathfrak{B}'$. Thus we may define $h : \partial\mathfrak{A} \to \partial\mathfrak{B}$ by $h = h''^{-1} \circ \tilde{f} \circ h'$. It is straightforward to show that h is an isomorphism which extends $\tilde{f}$; hence, it also extends f and the proof is complete. □

4 Natural numbers in Local Set Theories

The notion of natural numbers object in toposes has been widely studied in the literature. However, the original definition of natural numbers object, introduced by F.W. Lawvere in 1964 (see Lawvere (1964)), makes sense in any category with finite products.

In this section we briefly recall some elementary results on natural numbers in a local set theory, including a statement of the Cantor's back and forth theorem proved in Sbardellini & Coniglio (2006).

A local set theory N is said to be *naturalized* if its language has a distinguished type N, a closed term 0 (zero) of type N and an N-function s (successor) of type N × N satisfying the usual (second-order) Peano axioms. The N-set $U_N = \{x_N : \top\}$ will be denoted by $\mathbb{N}$ and called N-set of natural numbers. The elements of $\mathbb{N}$ will be denoted by m, n, n' etc.

In a naturalized local theory N the following *primitive recursion principle* (PRP) is valid (see Bell (1988, n.d.) for details):

$$x \in X, g \in X^{X \times \mathbb{N}} \vdash_N \exists! f \in X^{\mathbb{N}}[f(0) = x \;\wedge\; \forall n. f \circ s(n) = g(f(n), n)].$$

Consider the N-function $[\cdot] : \mathbb{N} \to P\mathbb{N}$ defined by PRP as follows:

$$\vdash_N \exists! [\cdot] \in (P\mathbb{N})^{\mathbb{N}} [[0] = \emptyset \;\wedge\; [s(n)] = [n] \cup \{n\}].$$

The relation $m < n :\Leftrightarrow m \in [n]$ is a strict order in $\mathbb{N}$ such that, as expected, $\vdash_N [n] = \{m : m < n\}$. Then we may show, as in the classical case, some basic facts about the natural numbers such as existence of minimal element (namely, 0), discretion, irreflexivity, transitivity, linearity and decidability, this last allowing us to define the linear partial order $\leq$ by $m \leq n :\Leftrightarrow (m = n) \vee (m < n)$. Using PRP it is straightforward to define addition, product and exponentiation as appropriate N-functions.

As usual, the integers object $\mathbb{Z}$ is defined as the coproduct of $\mathbb{N}$ with its image by s. Extending strict order, addition and product from the system $\mathbb{N}$, the resulting $\mathbf{C}(N)$-structure is a linearly ordered commutative ring. From this, the rational numbers object $\mathbb{Q}$ can be defined by reproducing the classical quotient construction.[3] Finally, extending strict order, addition and product from $\mathbb{Z}$, the resulting $\mathbf{C}(N)$-structure is again a linearly ordered commutative ring. Furthermore, the linearly ordered $\mathbf{C}(N)$-structure $\langle \mathbb{Q}, < \rangle$ is persistent (recall the end of Section 2).

We recall from Sbardellini & Coniglio (2006) the following notation: the finite conjunction of a formula α is the N-function $\bigwedge_{i<(\cdot)} \alpha : \mathbb{N} \to U_\Omega$ defined by PRP:

$$\mathrel{\vdash_N} \bigwedge_{i<0} \alpha = \top \ \wedge\ \forall n \left[\bigwedge_{i<n+1} \alpha = \bigwedge_{i<n} \alpha \ \wedge\ \alpha(n) \right].$$

Analogously, it is defined the finite disjunction $\bigvee_{i<(\cdot)} \alpha : \mathbb{N} \to U_\Omega$. It can be easily shown that, if the formula α is decidable in $\mathbb{N}$, then so are $\bigwedge_{i<n} \alpha$ and $\bigvee_{i<n} \alpha$ for each n (cf. Sbardellini & Coniglio (2006)).

The next useful result was proved in Sbardellini & Coniglio (2006).

Proposition 4.1 (Minimum principle). *If the formula α is decidable in $\mathbb{N}$, then*
$$\exists n.\alpha(n) \mathrel{\vdash_N} \exists n[\alpha(n) \ \wedge\ \forall m(\alpha(m) \to n \le m)].$$

Since $\le$ is antisymmetric, the minimum will be unique. This unique minimum will be denoted by $\mu n.\alpha(n)$.

A given N-set X is said to be *denumerable* if it satisfies the axiom $\vdash_N \mathrm{DEN}(X)$, where

$$\mathrm{DEN}(X) :\Leftrightarrow X = \emptyset \ \vee\ \exists g \in X^{\mathbb{N}}.\mathrm{SUR}(g)$$

and

$$\mathrm{SUR}(g) :\Leftrightarrow \forall y \exists x.g(x) = y.$$

An N-map g such that $\mathrm{SUR}(g)$ is said to be an *enumeration* of X, and will be frequently denoted by g_X. If X satisfies the property $\vdash_N X \simeq \mathbb{N}$, where $X \simeq Y$ is the obvious formula stating that there exists an isomorphism in $\mathbf{C}(N)$ between X and Y, we say that X is *completely denumerable* or simply *countable*. Clearly, if X and Y are denumerable (respectively countable) N-sets, then the product $X \times Y$ is denumerable (resp. countable). An N-set X is *finite* if satisfies $\vdash_N \mathrm{FIN}(X)$, where $\mathrm{FIN}(X) :\Leftrightarrow \exists n.[X \simeq [n]]$.

---

[3]The reader interested in the details of this construction can consult, for instance, MacLane & Moerdijk (1992).

A *sequence on* X is an N-map $f : \mathbb{N} \to X + \{\sharp\}$, where $+$ denotes the coproduct in $\mathbf{C}(N)$ and $\sharp$ is any element, say 0. Given that $X + \{\sharp\}$ represents a disjoint union in N, it is tacitly assumed that $\vdash_N \sharp \notin X$. A *finite sequence on* X is defined to be a sequence on X satisfying the axiom $\vdash_N \mathrm{Fsq}_X(f)$, where

$$\mathrm{Fsq}_X(f) :\Leftrightarrow \exists n [f([n]) \subseteq X \ \wedge \ f(\mathbb{N} - [n]) = \{\sharp\}].$$

It is worth noting that the natural number n in the above definition is unique.

The N-set of all the finite sequences of X is given by:

$$X^* := \{ f \in (X + \{\sharp\})^{\mathbb{N}} \ : \ \mathrm{Fsq}_X(f) \}.$$

The elements of X^* will be denoted by $\vec{x}$, $\vec{x}_0$, $\vec{x}_1$ etc.[4] The element of X^* given by the constant N-map $n \mapsto \sharp$, will be denoted by $\sharp$, by abuse of notation. It was shown in Sbardellini & Coniglio (2006) that, if the N-set X is denumerable (resp. countable), then X^* is denumerable (resp. countable).

We can associate, to each finite sequence $\vec{x}$ of X^*, a natural number which (intuitively) represents the length or number of relevant elements of $\vec{x}$. Thus, consider the N-map $\mathrm{LGH}_X : X^* \to \mathbb{N}$ given by $\vec{x} \mapsto \mu n.[\vec{x}(n) = \sharp]$. Using this N-map we can collect all the finite sequences of a given length:

$$X^n := \{ \vec{x} \in X^* \ : \ \mathrm{LGH}_X(\vec{x}) = n \}.$$

The *effective image* of a finite sequence $\vec{x}$ is defined as follows:

$$\mathrm{EIM}_X(\vec{x}) := \vec{x}([\mathrm{LGH}_X(\vec{x})]) = \{ \vec{x}(n) \ : \ n < \mathrm{LGH}_X(\vec{x}) \}.$$

From now on, and when no confusion arises, we will write Fsq, LGH and EIM instead of Fsq_X, LGH_X e EIM_X, respectively.

The following topos version of Cantor's back and forth theorem, as well as the technical Corollary 4.3, were proved in Sbardellini & Coniglio (2006).

Theorem 4.2 (Cantor's back and forth theorem). *Let* $\langle A, < \rangle$ *and* $\langle B, < \rangle$ *be linearly ordered, inhabited, linear, persistent and denumerable* $\mathbf{C}(N)$-*structures. Then* $\langle A, < \rangle \simeq \langle B, < \rangle$.

Corollary 4.3. *Suppose the following conditions:*

---

[4]Note that this notation has already been used for contexts. However, this ambiguity will not lead to confusion.

- $\langle A, < \rangle$ and $\langle B, < \rangle$ are linearly ordered, inhabited, linear, persistent and denumerable **C**(N)-structures;

- $\alpha(x, y)$ is a formula decidable in $A \times B$;

- $\{a \ : \ \exists b.\alpha(a, b)\}$ is persistent in B and $\{b \ : \ \exists a.\alpha(a, b)\}$ is persistent in A.

Then there exists an isomorphism $h \ : \ \langle A, < \rangle \to \langle B, < \rangle$ such that $\mathrel{\vdash\kern-0.6em\sim}_N$ $\forall a.\alpha(a, h(a))$.

5 Dedekind real numbers and homogeneity

The Dedekind real numbers object is the N-set of all the Dedekind cuts (recall Section 3) over the ordered structure $\mathfrak{Q} = \langle \mathbb{Q}, < \rangle$ of the rational numbers. Thus, the Dedekind real numbers object is given by:

$$\mathbb{R}_d := \eth\mathbb{Q};$$

and the corresponding ordered **C**(S)-structure is obtained as:

$$\mathfrak{R}_\eth := \eth\mathfrak{Q}.$$

Then all the general results established in section 3 for Dedekind cuts can be applied in particular to $\mathfrak{R}_\eth$. Our aim in this section is to prove that this structure is homogeneous.

Recall from Sbardellini & Coniglio (2006) that, if $\langle A, < \rangle$ is a (partially) ordered **C**(N)-structure, a finite sequence $f : \mathbb{N} \to A \times A$ preserves the pairing order if $\mathrel{\vdash\kern-0.6em\sim}_N \text{PPO}(f)$, where

$$\text{PPO}(f) :\Leftrightarrow \bigwedge_{m < n < \text{LGH}(f)} [\pi' \circ f(m) < \pi' \circ f(n) \leftrightarrow \pi'' \circ f(m) < \pi'' \circ f(n)].$$

An order $\langle A, < \rangle$ is said to be *homogeneous* if, for every finite sequence $f : \mathbb{N} \to A \times A$ which preserves the pairing order, there exists an automorphism of $\langle A, < \rangle$ which extends $\text{EIM}(f)$. Intuitively, every finite partial isomorphism can be extended to an automorphism. The order $\langle A, < \rangle$ is said to be *effectively* homogeneous if there exists an effective (constructively defined) procedure for extending every f (see Sbardellini & Coniglio (2006) and Sbardellini (2005) for details).

It was proved in Sbardellini & Coniglio (2006) that the structure $\mathfrak{Q}$ is effectively homogeneous. Moreover, it is the least effectively homogeneous structure (up to isomorphisms), in the sense that it can be constructed a

monomorphism from $\mathfrak{Q}$ into every other effectively homogeneous structure.

Before the proof of the homogeneity of $\mathfrak{R}_\mathfrak{d}$ (Theorem 5.2) we need a
technical lemma, which uses the N-map $* : X^* \times X \to X^*$ defined by:

$$\vdash_N \quad \forall n[(n = \text{LGH}(\vec{x}) \to (\vec{x} * x)(n) = x)$$
$$\wedge \ (n \neq \text{LGH}(\vec{x}) \to (\vec{x} * x)(n) = \vec{x}(n))].$$

Intuitively, $*$ adds a new element of X to any sequence in X^*. The usual
relation of *apartness* can be defined in a given partial order $\langle A, < \rangle$ as
follows:

$$a' \# a'' :\Leftrightarrow (a' < a'') \vee (a'' < a').$$

Using $*$ and $\#$, we obtain the following:

Lemma 5.1. *If $\vec{r}$ is a finite sequence in $\mathbb{R}_d$, then the N-set $\mathbb{Q}_{\vec{r}}$, defined by*

$$\mathbb{Q}_{\vec{r}} := \left\{ q \ : \ \bigwedge_{n < \text{LGH}(\vec{r})} \vec{r}(n) \# i(q) \right\},$$

is dense in $\mathbb{Q}$ and denumerable.

Proof. First we show, by induction on the length of $\vec{r}$, that

$$\vdash_N \forall \vec{q} \in \mathbb{Q}^* \ [\ (\text{LGH}(\vec{q}) = \text{LGH}(\vec{r}) + 1 \ \wedge \ \bigwedge_{n < m < \text{LGH}(\vec{q})} \vec{q}(n) \neq \vec{q}(m)) \\ \to \bigvee_{n < \text{LGH}(\vec{q})} \vec{q}(n) \in \mathbb{Q}_{\vec{r}}] \qquad (3)$$

(intuitively, every finite sequence of rational numbers with $\text{LGH}(\vec{r})+1$ differ-
ent elements has at least one element in $\mathbb{Q}_{\vec{r}}$). If $\text{LGH}(\vec{r}) = 0$, then $\mathbb{Q}_{\vec{r}} = \mathbb{Q}$ and
the checking is immediate. Now let $\text{LGH}(\vec{r}) = n+1$ (so that $\text{LGH}(\vec{q}) = n+2$):
in this case, if we take (the unique) $\vec{r}_0$ such that $\vec{r} = \vec{r}_0 * \vec{r}(\text{LGH}(\vec{r}) - 1)$, by
induction hypothesis we have:

$$\vdash_N \bigvee_{n,m < \text{LGH}(\vec{q})} [\ \vec{q}(n) \neq \vec{q}(m) \ \wedge \ \vec{q}(n), \vec{q}(m) \in \mathbb{Q}_{\vec{r}_0} \ ].$$

On the other hand, from the definition of Dedekind cut (recall the expression (1) at the beginning of Section 3) it follows that

$$\vdash_N \ (q(n) \in \pi'(\vec{r}(\text{LGH}(\vec{r}) - 1))) \ \vee \ (q(m) \in \pi''(\vec{r}(\text{LGH}(\vec{r}) - 1))).$$

Thus:

$$\vdash_N \bigvee_{n,m < \text{LGH}(\vec{q})} [\ \vec{q}(n) \neq \vec{q}(m) \ \wedge \ (\vec{q}(n) \in \mathbb{Q}_{\vec{r}} \vee \vec{q}(m) \in \mathbb{Q}_{\vec{r}}) \ ]$$

and the expression (3) is proved. So we can take the element $q_0 \in \mathbb{Q}_{\vec{r}}$ defined by:

$$q_0 := \min\{q \ : \ g_{\mathbb{Q}}(q) < \text{LGH}(\vec{r} + 1)\},$$

where the above minimum has the obvious meaning, and construct a surjective map $g : \mathbb{Q}^{\text{LGH}(\vec{r})+1} \twoheadrightarrow \mathbb{Q}_{\vec{r}}$ by:

$$\vdash_N \left[\bigwedge_{n < m < \text{LGH}(\vec{q})} \vec{q}(n) \neq \vec{q}(m) \ \rightarrow \ g(\vec{q}) = \min(\vec{q}) \right]$$
$$\wedge \left[\neg \bigwedge_{n < m < \text{LGH}(\vec{q})} \vec{q}(n) \neq \vec{q}(m) \ \rightarrow \ g(\vec{q}) = q_0 \right].$$

Since $\mathbb{Q}^{\text{LGH}(\vec{r})+1}$ is totally denumerable, we conclude that $\mathbb{Q}_{\vec{r}}$ is denumerable. The density of $\mathbb{Q}_{\vec{r}}$ in $\mathbb{Q}$ follows from the expression bellow, which says intuitively that, given two rational numbers, there are arbitrarily many others between them:

$$\vdash_N \forall q', q'' \exists \vec{q} \left[\vec{q}(0) = \frac{q' + q''}{2} \ \wedge \ \bigwedge_{n < \text{LGH}(\vec{r})+1} \vec{q}(n+1) = \frac{\vec{q}(n) + q''}{2} \right].$$

Hence, it suffices to apply expression (3). $\qquad\square$

Theorem 5.2. *The $\mathbf{C}(N)$-structure $\mathfrak{R}_{\mathfrak{d}} = \langle \mathbb{R}_d, < \rangle$ of the ordered Dedekind real numbers object is (effectively) homogeneous.*

Proof. Let $f : \mathbb{N} \rightarrow \mathbb{R}_d \times \mathbb{R}_d$ be a finite sequence preserving the pairing order. By Lemma 5.1, the N-sets $\mathbb{Q}_{\pi' \circ f}$ and $\mathbb{Q}_{\pi'' \circ f}$ are denumerable and dense in $\mathbb{Q}$, and hence they are also persistent. Now let α be the formula defined by:

$$\alpha(q', q'') :\Leftrightarrow \bigwedge_{n < \text{LGH}(f)} \left[\pi' \circ f(n) < q' \leftrightarrow \pi'' \circ f(n) < q'' \right].$$

Observe that α is decidable in $\mathbb{Q}_{\pi' \circ f} \times \mathbb{Q}_{\pi'' \circ f}$ and, moreover,

$$\vdash_N \{q' \in \mathbb{Q}_{\pi' \circ f} \ : \ \exists q'' \in \mathbb{Q}_{\pi'' \circ f} . \alpha(q', q'')\} = \mathbb{Q}_{\pi' \circ f},$$

as well as

$$\vdash_N \{q'' \in \mathbb{Q}_{\pi'' \circ f} \ : \ \exists q' \in \mathbb{Q}_{\pi' \circ f} . \alpha(q', q'')\} = \mathbb{Q}_{\pi'' \circ f}.$$

In effect, by induction on the length of f we easily prove that

$$\vdash_N \forall q' \in \mathbb{Q}_{\pi' \circ f} \exists q'' \in \mathbb{Q}_{\pi'' \circ f} . \alpha(q', q'')$$

and also the converse. So, by Corollary 4.3 there exists an isomorphism $h : \langle \mathbb{Q}_{\pi' \circ f}, < \rangle \rightarrow \langle \mathbb{Q}_{\pi'' \circ f}, < \rangle$ such that $\vdash_N \forall q \in \mathbb{Q}_{\pi' \circ f} . \alpha(q, h(q))$ (that is, h extends $\text{EIM}(f)$). Now, by Lemma 3.6, there exists an automorphism $\tilde{h}$ of $\mathfrak{R}_{\mathfrak{d}}$ which extends h and therefore also extends $\text{EIM}(f)$. $\qquad\square$

Finally, it should be observed that, with slight modifications, the proof above can be adapted to the Cauchy real numbers defined in any topos with natural numbers object. That is:

Theorem 5.3. *The* $C(N)$*-structure* $\mathfrak{R}_c = \langle \mathbb{R}_c, < \rangle$ *of the ordered Cauchy real numbers object is (effectively) homogeneous.*

A detailed construction of the Cauchy real numbers defined in local set theories, as well as several missing details of the proof of Veldman and Waaldijk's result, can be found in Sbardellini (2005).

References

Bell, J. (1988), *Toposes and local set theories*, Oxford University Press.

Bell, J. (n.d.), The development of categorical logic, Preprint. Available at URL = http://publish.uwo.ca/~jbell/catlogprime.pdf.

Bell, J. & Slomson, A. (1971), *Models and ultraproducts: an introduction*, North-Holland, Amsterdam.

Brouwer, L. (1907), Over de grondslagen der wiskunde (*'On the Foundations of Mathematics', in German*), PhD thesis, University of Amsterdam, Department of Physics and Mathematics, Amsterdam.

Constable, R. & Bickford, M. (2014), 'Intuitionistic completeness of first-order logic', *Annals of Pure and Applied Logic* **165**(1), 164–198.

de Swart, H. (1978), 'First steps in intuitionistic model theory', *The Journal of Symbolic Logic* **43**(1), 3–12.

Eilenberg, S. & MacLane, S. (1945), 'General theory of natural equivalences', *Transactions of the American Mathematical Society* **58**, 231–294.

Fitting, M. (1969), *Intuitionistic Logic, Model Theory and Forcing*, North-Holland.

Johnstone, P. (1977), *Topos theory*, Academic Press, London.

Jónsson, B. (1960), 'Homogeneous universal relational systems', *Mathematica Scandinavica* **8**, 137–142.

Lawvere, F. (1964), 'An elementary theory of the category of sets', *Proceedings of the National Academy of Sciences* **52**, 1506–1511.

MacLane, S. & Moerdijk, I. (1992), *Sheaves in geometry and logic: a first introduction to topos theory*, Springer-Verlag, New York.

Mulvey, C. (1974), Intuitionistic algebra and representation of rings, *in* J. Liukkonen & K. Hofmann, eds, 'Recent Advances in the Representation of Rings and C*-Algebras by Continuous Sections', Vol. 148 of *Memoirs of the AMS*, AMS, Providence, RI, pp. 3–57.

Sbardellini, L. (2005), O Continuum, os Reais e o Conceito de Homogeneidade *('The Continuum, the Real Numbers and the Concept of Homogeneity', in Portuguese)*, PhD thesis, State University of Campinas, Department of Philosophy - IFCH, Campinas, Brazil.

Sbardellini, L. & Coniglio, M. (2006), 'Some results on ordered structures in toposes', *Reports on Mathematical Logic* **40**, 183–200.

Veldman, W. & Waaldijk, F. (1996), 'Some elementary results in intuitionistic model theory', *The Journal of Symbolic Logic* **61**(3), 745–767.

The True, the False, and the slingshot arguments

Oswaldo Chateaubriand*

* Departamento de Filosofia
Pontifícia Universidade Católica do Rio de Janeiro
ochateaubriand@gmail.com

It is a great pleasure for me to contribute to this Festschrift in honor of Luiz Carlos Pereira's 60th birthday. We have been colleagues and friends for over thirty years and have had an intense academic and intellectual interaction. Of the many logical and philosophical subjects we have discussed, the problem of the denotation of sentences has been of great importance in my own work, and in this paper I will reconsider this issue.

1 A general principle about meaning and denotation

When Frege reformulated his concept script in the early 1890's he postulated two logical objects as the referents of true and false sentences. He called these objects 'the True' and 'the False' and argued that all true sentences referred to the True and all false sentences referred to the False.[1] Frege's arguments were not entirely convincing, and in the 20th century several logicians and philosophers offered formal arguments for Frege's conclusion, and these came to be known collectively as "the slingshot arguments". In this paper I will present a simple refutation of three main arguments — given by Church, Davidson, and Gödel—based on a general principle about meaning and denotation.[2]

Consider a relation Ξ among sentences which does not preserve truth-value; i.e., such that for sentences R and S we may have

$$R \equiv S$$

---

[1] Frege introduces these logical objects, and his re-interpretation of the concept script, in "Function and Concept", and argues at length for their role as the denotation of sentences in "On Sense and Reference". I will use 'sentence' throughout the paper, although different authors use 'thought', 'proposition', and 'statement', among others.

[2] These arguments are discussed in great detail in Chapter 4 of Logical Forms and their rejection was a main motivation for the developments of the views presented in the book. Frege's arguments are also discussed in detail in Chapter 2

with R and S having different truth-values — where by truth-value I mean true, false, or neither true nor false. The general principle (**GP**) is that the fact that two sentences have the relation Ξ does not support the conclusion that these sentences have the same denotation (or the same meaning). The justification for this principle is quite obvious, because if the sentence R denotes the True or the False and S denotes the False or the True or does not denote, then R and S don't have the same denotation.[3]

An important example of such a relation is *logical equivalence*, in the sense of mutual logical consequence. If two sentences are logically equivalent and one of them is true, the other must also be true, but we can have logically equivalent sentences where one is false and the other is neither true nor false. As I have emphasized on several occasions[4] Frege's and Russell's analyses of sentences of the form

(1) the F is G

are logically equivalent but do not have always the same truth-value. For, according to Frege, (1) is true if, and only if

(1f) 'the F' denotes and its denotation is G,

but if there isn't a unique F, then on Frege's analysis (1) is truth-valueless — because 'the F' does not denote — whereas on Russell's analysis it is false, because it means

(1r) There is a unique thing that is F and that thing is G.

Yet in both cases (1) is true if, and only if, there is a unique F that is G, and therefore (1f) and (1r) are logical consequences of each other.

Let me now turn to the three slingshot arguments to which I referred at the beginning. Two main principles that are used in all these arguments are Frege's principle about definite descriptions as singular terms

(D) If there is a unique object that satisfies the property (or condition) F, then the definite description 'the F' denotes this object; otherwise 'the F' does not denote.
and Frege's substitutivity principle

(R) The denotation of a composite expression depends only on the denotation of its parts and not on their sense.

---

[3]Although Church adheres explicitly to Frege's postulation of the objects the True and the False, both Davidson and Gödel offer their arguments hypothetically: if sentences denote, then all true sentences denote the same thing and all false sentences denote the same thing.

[4]See, for example, Chapter 3 of *Logical Forms* and "Descriptions: Frege and Russell Combined"

2 Davidson's argument[5]

My rejection of Davidson's argument in Chapter 4 of *Logical Forms* was a direct application of principle (GP) for logical equivalence, which Davidson assumed to preserve denotation. His argument consists of the following steps, where R and S are sentences that have the same truth-value.

(1) R

(2) $\{x \colon x = x \& R\} = \{x \colon x = x\}$

(3) $\{x \colon x = x \& S\} = \{x \colon x = x\}$

(4) S

The steps from (1) to (2) and from (3) to (4) are justified by Davidson's assumption that logical equivalence preserves denotation, because these pairs are indeed logically equivalent. The step from (2) to (3) is justified by principle (R), since given that R and S have the same truth-value, then the set-terms on the left-hand-side of (2) and (3) denote the same set, universal or empty, depending on whether R and S are both true or both false. What I did not realize at the time was that exactly the same considerations I used to reject the steps from (1) to (2) and from (3) to (4) can be applied to the arguments given by Church and by Gödel, as I will show in what follows.

3 Church's argument[6]

Church argues that two sentences that appear to have nothing in common aside from their truth-value have the same denotation. The argument consists of the following steps:

(1) Sir Walter Scott is (=) the author of *Waverley*.

(2) Sir Walter Scott is (=) the man who wrote twenty-nine *Waverley* novels altogether.

(3) The number, such that Sir Walter Scott wrote that many *Waverley* novels altogether, is (=) twenty-nine.

(4) The number of counties in Utah is (=) twenty-nine.

---

[5]In "Truth and Meaning", pp. 19.

[6]In *Introduction to Mathematical Logic* 1, pp. 24-25.

These statements are interpreted as identities, and to emphasize this I included the identity sign in parentheses.

The transitions from (1) to (2) and from (3) to (4) are justified by principles (R) and (D). The problem for Church is to justify the transition from (2) to (3), and his argument is that if these two sentences are not actually synonymous, they are so nearly so as to ensure that they have the same denotation. My reason for rejecting this conclusion is now the following. Consider the pair

(2*) Sir Walter Scott is (=) the man who wrote thirty *Waverley* novels altogether.

and

(3*) The number, such that Sir Walter Scott wrote that many *Waverley* novels altogether, is (=) thirty.

(2*) and (3*) are as close in meaning as are (2) and (3), and by Church's argument they should also be synonymous, or nearly synonymous, *and have the same denotation*. It so happens, however, that (2*) is neither true nor false, because the definite description 'the man who wrote thirty *Waverley* novels altogether' does not denote, whereas (3*) is false, because the number of *Waverley* novels Sir Walter Scott wrote is twenty-nine, not thirty. Hence, by principle (GP), the alleged synonymy, or near synonymy, of (2) and (3) is illusory, and does not guarantee that they have the same denotation.

4   Gödel' s argument[7]

In its full generality Gödel's argument depends on several additional principles, and the way I reconstructed it goes as follows:

(1) R

(2) Fa

(3) $a = \iota x(Fx \& x = a)$

(4) $a = \iota x(x = a \& b = b \& x = a)$

(5) $a = a \& b = b$

(6) $b = \iota x(a = a \& x = b \& x = b)$

---

[7]In "Russell's Mathematical Logic", p. 214 note 5.

(7) $b = \iota x(Gx \& x = b)$

(8) Gb

(9) S

The transitions from (1) to (2) and from (8) to (9) are justified by a claim that any sentence can be put in the form of a predication. The transitions from (3) to (4), (4) to (5) and (5) to (6) are justified by principles (R) and (D) — although some questions can be raised about this.

The crux of the matter, however, is the justification of the transitions from (2) to (3) and from (7) to (8), for which Gödel appeals to the following principle:

(G1)   A sentence of the form 'Fa' means the same thing (and has the same denotation) as a sentence of the form '$a = \iota x(Fx \& x = a)$'.

Although it is clear that this principle does not hold for false sentences of the form 'Fa' — because in this case the descriptive term '$\iota x(Fx \& x = a)$' does not denote, and hence the identity is neither true nor false — all the analyses I know restrict the argument with the assumption that Fa is true.[8] This is a direct violation of my general principle (GP), because since a false sentence cannot mean the same (and have the same denotation) as a truth-valueless sentence, the principle (G) cannot be assumed to hold for true sentences either. Although Gödel's principle, as he stated it originally, seems very plausible, I have an explanation for this plausibility that uses a different formalization — but this requires a detour to my analysis of descriptive terms and descriptive predicates.

5   Descriptive terms and descriptive predicates

By a *descriptive term* I understand a singular term of the form 'the so-and-so', such as 'the present King of France', 'the author of *Waverley*', and so on. Most descriptive terms are obtained from descriptive functions filling up argument places. Thus, the term 'the author of Waverley' can be obtained from the descriptive function 'the author of x' substituting '*Waverley*' for 'x'. The basic feature of a descriptive term is given by principle (D).

By a *descriptive predicate* I understand a predicate of the form 'is the so-and- so', such as 'is the present King of France', 'is the author of

---
[8]I do this myself in Chapter 4 of *Logical Forms*, pp. 147. I think this is my only error in that chapter's discussion of the slingshot arguments. The detailed discussion of Gödel's argument is on pp. 146–153.

Waverley', and so on. A descriptive predicate of the form 'x is the F' is obtained from the predicate 'x is an F' by adding a uniqueness clause 'and only x is an F' — or 'and nothing other than x is an F'. That is, 'x is the F' is the predicate

$$[Fx \ \& \ \forall y(Fy \to y = x)](x).$$

It is important to realize that descriptive predicates are part of quantification logic, and do not require additional interpretation. In other words, they are available to users of quantification logic independently of any views about descriptive terms and descriptive functions.

6 Gödel's argument continued

Gödel's principle (G1) is actually formulated as follows:

(G) 'Fa' and the sentence 'a is the object that has the property F and is identical to a' mean the same thing.

Although the interpretation of (G) is taken to be (G1) — by Gödel and everyone else — according to my distinction between descriptive terms and descriptive predicates, the right hand side of (G) can also be interpreted as the predication

[x is the object that has the property F and is identical to a](a)

i.e., formally,

$$[Fx \ \& \ \forall y(Fy \to y = x)\& x = a](a).$$

Now (G) can be formulated as

(G2) A sentence of the form 'Fa' means the same thing (and has the same denotation) of the form '$[Fx \ \& \ \forall y(Fy \to y = x)\& x = a](a)$'.

The interesting thing about this analysis is that if Fa is false the right side of (G2) is also false, and thus we do not have a difference of truth-value. We can conclude, therefore, that whereas the right hand sides of (G1) and (G2) are logically equivalent, they do not have the same meaning, and that the proper formulation of Gödel's principle is (G2). But, of course, now Gödel's argument does not go through.[9]

---

[9]Gödel himself points this out on pp. 213 for the interpretation of (G) using Russell's analysis of descriptions.

References

Chateaubriand, O. (2001), *Logical Forms. Part I: Truth and Description*, UNICAMP.

Chateaubriand, O. (2002), 'Descriptions: Frege and Russell Combined', *Synthese* **130**, 213–226.

Church, A. (1956), *Introduction to Mathematical Logic 1*, Princeton.

Davidson, D. (1967), *Inquiries into Truth and Interpretation*, Clarendon Press, chapter Truth and Meaning.

Frege, G. (1891), *Geach and Black Translations from the Philosophical Writings of Gottlob Frege*, Blackwell's, chapter Function and Concept.

Frege, G. (1892), *Geach and Black Translations from the Philosophical Writings of Gottlob Frege*, Blackwell's, chapter On Sense and Reference.

Gödel, K. (1944), *Benacerraf and Putnam Philosophy of Mathematics: Selected Readings*, Prentice Hall, chapter Russell's Mathematical Logic.

On being naturally general

P. A. S. Veloso*   S. R. M. Veloso[†]   L. B. Vana[‡]

* Systems and Computer Engineering Program, COPPE
Universidade Federal do Rio de Janeiro
pasveloso@gmail.com

[†] Department of Systems and Computer Engineering
Universidade Estadual do Rio de Janeiro
sheila.murgel.bridge@gmail.com

[‡] Institute of Mathematics
Universidade Federal Fluminense
leobvana@gmail.com

Abstract

We examine the structure of natural deduction derivations for some versions of 'generally' represented by generalized quantifiers. Versions of 'generally' (e.g. 'most', 'many', 'several') appear often in ordinary language and in some branches of science. To handle assertions with such vague notions, First-Order Logic is extended to logics for 'generally', which can be formulated as natural deduction systems with some special rules. These natural deduction systems for 'generally' share some macroscopic properties with First-Order Logic, but may differ with respect to some microscopic properties. We introduce some special systems and analyze the inner structure of their natural deduction derivations.

1 Introduction

We will examine the structure of natural deduction derivations for some versions of 'generally' (represented by generalized quantifiers).

Vague notions (such as 'generally', 'rarely', 'most', 'many', 'several', 'few', etc.) appear often in assertions and arguments in ordinary language and in some branches of science (Barwise & Cooper, 1981; Veloso & Carnielli, 2004). One can provide logics that capture distinct intuitive notions of 'generally'. Logics for 'generally' (LG's) were introduced as extensions of First-Order Logic (FOL) for handling assertions with vague notions, like 'generally', 'most', 'many' and 'several', expressed by means of generalized formulas. Deductive systems have been developed for such LG's (Carnielli & Veloso, 1997;

Veloso & Veloso, 2001, 2002, 2004; Veloso & Carnielli, 2004; Veloso & Veloso, 2005a,b).

These deductive systems for LG's are macroscopically quite similar to those for FOL: they are sound, complete and have interpolation (Veloso, 2001; Veloso & Veloso, 2004). In particular, with derivation rules reflecting properties of 'generally', we can construct natural deduction systems for LG's that are normalizable (Vana et al., 2005, 2007). Other microscopic properties, however, differ from those for FOL.

Normal derivations in FOL have a familiar "hour-glass" form, consisting of an elimination region and introduction region separated by minimal formulas (Prawitz, 1965; Dalen, 1989). A normal path from hypothesis H to conclusion C uses elimination rules ($\searrow$) to reach a minimal formula B and then introduction rules ($\nearrow$) to obtain the conclusion; its structure can be represented as shown in Fig. 1(a). As a result, FOL has the sub-formula property: every formula in a normal derivation is a sub-formula of a hypothesis or of the conclusion.

LG's handle generalized information and the structure becomes as in Fig. 1(b). We now have a frontier region manipulating 'generally' (represented by a quantifier), possibly with several applications of transformation rules. In (Vana et al., 2014), we have characterized the inner structure of this frontier region for some specific LG's, roughly as generalized introductions ($\nearrow$) and eliminations ($\searrow$), separated by a single application of the transformation rule ($\updownarrow$), as shown in Fig. 1(c).

Figure 1: Derivation structure

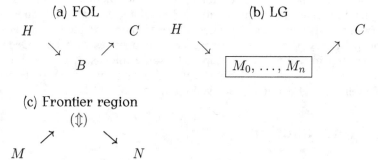

(a) FOL

H C H

(b) LG

     C

B

$M_0, \ldots, M_n$

(c) Frontier region

($\updownarrow$)

M N

Here, we will refine this result by examining the inner structure of natural deduction derivations for some versions of 'generally'. A motivation for this analysis is the "natural" way to establish that a set is in a filter: by showing it includes an intersection of finitely many sets known to be in the filter.

The structure of this paper is as follows. In Sect. 2, we briefly review some ideas about reasoning with 'generally' and corresponding logics. In Sect. 3, we examine natural deduction systems for 'generally'. In Sect. 4, we analyze the inner structure of natural deduction derivations. Finally, Sect. 5 presents some concluding remarks.

2 Reasoning with 'Generally'

We now briefly review some ideas about 'generally': reasoning and logics.[1]

We examine families for 'generally' (in 2.1) and corresponding logics (in 2.2).

2.1 Families for 'Generally'

We now introduce families for notions of 'generally'.

The diverse notions of 'generally' may have distinct behaviors.

Example 1 (Notions of 'generally'). *First, consider the universe of Brazilians and imagine that one accepts the two assertions:*
  (α): *"Several Brazilians shave their legs"*
  (β): *"Several Brazilians shave their faces".*
In this case, one is likely to accept also the assertion:

($\sqcup$) *"Several Brazilians have their faces shaved or sport a moustache";*

but one is not likely to accept the following assertion:

($\sqcap$) *"Several Brazilians shave their legs and their faces".*

Next, consider the universe of natural numbers and imagine that one accepts the two assertions:
  (γ): *"Most naturals are larger than 15"*
  (δ): *"Most naturals do not divide 12".*
Then, one would probably accept also the following two assertions:

($\vee$) *"Most naturals are larger than 15 or even", and*

($\wedge$) *"Most naturals are larger than 15 and do not divide 12".*

One may explain "Objects generally have property P" by saying that "the objects having property P form an important set". What sets are to be considered 'important' depends on the notion of 'generally', which may affect whether one accepts an argument or not.[2]

Example 2 (Argument with 'many'). *Consider the following 3 assertions.*

(v) *"Whoever likes sports watches channel SportTV" (precise premise).*

---

[1] For more motivation and details, see e. g. (Veloso, 2001; Veloso & Carnielli, 2004; Veloso & Veloso, 2011)

[2] In Example 1, we infer ($\sqcup$) from $\{(\alpha), (\beta)\}$, without obtaining ($\sqcap$), whereas we also conclude ($\vee$) and ($\wedge$) from $\{(\gamma), (\delta)\}$, even though ($\sqcap$) and ($\wedge$) have similar syntactic structures.

(σ) "Many *boys like sports*" *(vague premise).*

(τ) "Many *boys watch channel SportTV*" *(vague conclusion).*

The inference from (v) and (σ) to (τ) appears to be reasonable.

We now take a closer look at some diverse notions of 'generally',

By a *complex* over a universe M, we mean a family $\mathcal{K}$ of subsets of M (those considered important). A *module* amounts to a class of complexes.

Besides the *basic module* $\mathcal{B}$ (complexes without restriction), we one can also consider some specific modules, given by their properties. Table 1 shows some characteristic properties of complexes. In principle, each combination of such properties can be used to define a notion of 'generally'. Some of these modules are familiar, among these we can mention those given in Table 2.

Other equivalent descriptions for some modules are possible.[3] So, we have a hierarchy of modules as indicated in Fig. 2.

Table 1: Properties of complexes

Universe	(M)	$M \in \mathcal{K}$
Non-void	$(\emptyset)$	$\emptyset \notin \mathcal{K}$
Intersection-closed	$(\sqcap)$	$S, T \in \mathcal{K} \Rightarrow S \cap T \in \mathcal{K}$
Union-closed	$(\sqcup)$	$S, T \in \mathcal{K} \Rightarrow S \cup T \in \mathcal{K}$
Superset	$(\sqsupseteq)$	$S \in \mathcal{K} \& S \subseteq T \Rightarrow T \in \mathcal{K}$
Complement attracting	$(\neg)$	$S \notin \mathcal{K} \Rightarrow \overline{S} \in \mathcal{K}$
Complement repelling	$(\not\neg)$	$S \in \mathcal{K} \Rightarrow \overline{S} \notin \mathcal{K}$
Prime	$(\not\sqcup)$	$S \cup T \in \mathcal{K} \Rightarrow S \in \mathcal{K} \text{ or } T \in \mathcal{K}$
Co-intersection	$(\overrightarrow{\sqcap})$	$S \cap T \in \mathcal{K} \Rightarrow S \in \mathcal{K} \text{ and } T \in \mathcal{K}$

2.2 Logics for 'Generally'

We now indicate how one can formulate logics for 'generally'.

To express "objects generally have a given property", we add to First Order Logic (FOL) a new (variable binding) quantifier ∇ to represent 'generally', thus extending FOL language L to ∇-language L^{∇}.

The notions of variable occurring (free) in a formula are as usual. We shall use the notations $\text{oc}(F)$ for the set of variables occurring in formula F and $\text{fr}[\Gamma]$ for the set of variables with free occurrences in

---

[3]Superset ($\sqsupseteq$) can be obtained from co-intersection ($\overrightarrow{\sqcap}$), since $S \subseteq T$ iff $S = S \cap T$. Also, Union-closed ($\sqcup$) can be obtained from Superset ($\sqsupseteq$), as $S \subseteq T$ iff $S \cup T = T$.

some formula in the set Γ of formulas. We also use familiar notions like variable substitution ($F[v/t]$) and rank ($\mathrm{rk}(F)$).

Table 2: Modules and characteristic properties

($\mathcal{P}$) *Proper*: Universe (M) and Non-void ($\emptyset$.

($\mathcal{S}$) *Proper Up-closed*: Universe (M), Non-void ($\emptyset$ and Superset ($\supseteq$).

($\mathcal{L}$) *Proper Lattices*: (M), ($\emptyset$, closed under Intersection ($\sqcap$) and Union ($\sqcup$).

($\mathcal{F}$) *Proper Filters*: (M), ($\emptyset$, Intersection-closed ($\sqcap$) and Superset ($\supseteq$).

($\mathcal{U}$) *Proper Ultrafilters*: (M), ($\emptyset$, ($\sqcap$), ($\supseteq$) and Complement attracting ($^-$).

Figure 2: Hierarchy of modules

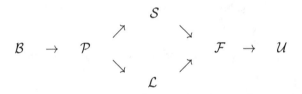

As mentioned, an assertion such as "Objects generally have property P" may be understood as "the set of objects that have the property P is important" (among the subsets of the universe of discourse). So, one gives the semantics for 'generally' by adding families of sets (those that are considered important) to a usual first-order structure and extending the definition of satisfaction to ∇.[4]

A *modulated structure* $\mathfrak{M}^{\mathcal{K}} = \langle \mathfrak{M}, \mathcal{K} \rangle$ consists of a usual structure $\mathfrak{M}$ together with a *complex* $\mathcal{K}$: a family of subsets of the universe M of $\mathfrak{M}$. We extend the definition of *satisfaction* of a formula in a structure under an assignment $s : V \to M$ to the variables as follows: for a generalized formula $\nabla v\, A$ we define

$$\mathfrak{M}^{\mathcal{K}} \models \nabla v\, A[\mathbf{s}] \text{ iff } \{b \in M : \mathfrak{M}^{\mathcal{K}} \models A[\mathbf{s}(v \mapsto b)]\}$$

belongs to the complex $\mathcal{K}$.[5]

---

[4]Note that our ∇ quantifier is not a Mostowski's quantifier (Mostowski, 1957).

[5]The updated assignment $\mathbf{s}(v \mapsto b)$ is as usual: it agrees with s on every variable, except v.

Other concepts, such as *model* ($\mathfrak{M}^{\mathcal{K}} \models A$ and $\mathfrak{M}^{\mathcal{K}} \models \Gamma$), are as usual.

On the other hand, the concept of consequence depends on the specific notion of 'generally' involved. For instance, the argument in Example 2 will be correct if the complexes are closed under supersets (which seems reasonable in the case of 'many'). More precisely, we say that a formula A is an *up-closed consequence* of a set Γ of sentences iff $\mathfrak{M}^{\mathcal{K}} \models A$ whenever $\mathfrak{M}^{\mathcal{K}} \models \Gamma$, for every model $\mathfrak{M}^{\mathcal{K}}$ whose complex $\mathcal{K}$ is closed under supersets (of its universe). We use the notation $\Gamma \models_{\mathcal{S}} A$, where $\mathcal{S}$ is the class of up-closed complexes.

In this manner, each notion of 'generally' gives rise to a corresponding consequence relation: $\Gamma \models_{\mathcal{C}} A$, where $\mathcal{C}$ is a given module, i. e. a class of complexes.[6]

We can axiomatize some logics for 'generally' by adding appropriate axiom schemes to an axiomatization for FOL.

We have the two basic *extensionality principles* given in Table 3.

Table 3: Basic extensionality schemes: equivalence and alphabetic variant

$[\leftrightarrow \nabla]$: $\forall v(A \leftrightarrow B) \to (\nabla v A \to \nabla v B)$

$[\nabla^{\alpha}]$: $\nabla v A \to \nabla u A^{[v/u]}$, if $v \notin o\alpha(A)$

The two valid schemes shown in Table 3 axiomatize the logic BL of the basic module $\mathcal{B}$. For some other modules, we also can axiomatize their properties.

For proper modules, we can use schemes or axioms as shown in Table 4.[7]

Table 4: Proper schemes and axioms (for Universe and Non-void)

	Schemes	Axioms
Universe	$[\forall \nabla] : \forall v\, A \to \nabla v\, A$	$[\nabla\top] : \nabla v \top$
Non-void	$[\not\exists \not\nabla] : \neg \exists v \neg A \to \neg \nabla v \neg A$	$[\not\nabla\bot] : \neg \nabla v \bot$

Other properties can be axiomatized by schemes like those in Table 5.

Thus, we can axiomatize logics like those of the modules mentioned in 2.1: Families for 'Generally' (cf. Table 2). Note that $[\leftrightarrow \nabla] \vdash [\wedge \nabla] \leftrightarrow [\to \nabla]$. Also, in some of the schemes and axioms in Tables 4 and 5 we can recognize properties of the generalized quantifier ∇ similar to those of the usual quantifiers $\forall$ and $\exists$. Notice, however, that consecutive ∇'s do not commute (cf. (Veloso & Veloso, 2004), p. 158).

---

[6]These consequence relations are extensional: $\Gamma \models_{\mathcal{C}} \forall v\, (A \leftrightarrow B) \to (\nabla v\, A \to \nabla v\, B)$.

[7]We use $\top$ as an abbreviation for $\bot \to \bot$. Note that $\vdash \top$. We use $\neg A$ as an abbreviation for $A \to \bot$. For $[\not\exists \not\nabla]$, we can also use simpler equivalent formulations, depending on the underlying FOL: $\neg \exists v\, A \to \neg \nabla v\, A$, for the intuitionistic case, and $\nabla v\, A \to \exists v\, A$, for the classical case.

Table 5: Specific schemes

Intersection-closed	$[\nabla\wedge]$	$(\nabla v A \wedge \nabla v B) \to \nabla v (A \wedge B)$
Union-closed	$[\nabla\vee]$	$(\nabla v A \wedge \nabla v B) \to \nabla v (A \vee B)$
Superset	$[\to \nabla]$	$\forall v(A \to B) \to (\nabla v A \to \nabla v B)$
Complement attracting	$[\nabla\neg]$	$\neg\nabla v A \to \nabla v \neg A$
Complement repelling	$[\neg\nabla]$	$\nabla v \neg A \to \neg\nabla v A$
Prime	$[\vee\nabla]$	$\nabla v (A \vee B) \to (\nabla v A \vee \nabla v B)$
Co-intersection	$[\wedge\nabla]$	$\nabla v (A \wedge B) \to (\nabla v A \wedge \nabla v B)$

3   Natural Deduction for 'Generally'

We now examine natural deduction for 'generally' (Vana et al., 2005, 2007; Vana, 2008). We use familiar concepts, like track and order, and notations (Prawitz, 1965; Dalen, 1989).

We will extend a natural deduction system for FOL to handle 'generally'. The expected structure of a derivation is a FOL derivation with local manipulations of ∇. It will be convenient to employ marked formulas.

We first introduce marked formulas and then examine natural deduction systems for some logics of 'generally' (in 3.1 and 3.2).

We now introduce marked formulas to discipline manipulations of ∇.

Some reasons for using marked formulas are as follows. On the one hand, it is convenient to view a generalized formula $\nabla v\, A(v)$ as indecomposable.[8] On the other hand, the interplay between the generalized quantifier ∇ and other logical constants, such as the propositional connectives and the usual universal and existential quantifiers, depends on the specific LG at hand; marked formulas turn out to be useful for expressing such interplay. So, marked formulas play an important role, highlighting the regions where generalized information is manipulated.

Marked formulas have the same intended meaning as generalized formulas. A marked formula has the form $\langle A(\_)\rangle$ and is intended to represent generalized formula $\nabla v\, A(v)$. The idea is that '$\_$' represents a "generic object" while '$\langle$' and '$\rangle$' emphasize that $A(v)$ is the scope of the generalized quantifier. A marked formula may have occurrences of ∇.[9] We consider $\langle A(\_)\rangle$ as a sub-formula of $\nabla v\, A(v)$, with rank

---

[8]Notice that instantiation of generalized variables does not hold in LG's.

[9]For instance, $\nabla x p(x, \_)$ and $\nabla y p(\_, y)$ represent $\nabla y \nabla x p(x, y)$ and $\nabla x \nabla y p(x, y)$, respectively, wheras $p(\_, \_)$ represents $\nabla z p(z, z)$. See also Example 5 (Basic derivation).

$\text{rk}(\langle A(\_)\rangle) = \text{rk}(A)) + 0.5.$[10] We also consider other sub-formulas as follows: $\langle A(\_)\rangle$ is a sub-formula of $\langle\neg A(\_)\rangle$ and, for a binary connective $\star \in \{\wedge, \vee, \rightarrow, \leftrightarrow\}$, $\langle A(\_)\rangle$ and $\langle B(\_)\rangle$ are sub-formulas of $\langle A(\_) \star B(\_)\rangle$.[11]

More precisely, we consider a new symbol _ (not in L^∇) and create new formulas as follows. A *generic instance* of formula A with respect to variable v is the result $A[v/\_]$ of replacing every ccurrence, if any, of v in A by the new symbol _. A *marked formula* has the form $\langle A_\_\rangle$, where $A_\_$ is a generic instance.[12] We thus have the *marked fragment* MF, consisting of the marked formulas, and the *extended fragment* $EF = L^\nabla \cup MF$.

Consider a natural deduction system for FOL.[13] Since we wish to have marked formulas, we extend its rules to cover them. More specifically, we extend the elimination rules for $\vee$ and $\exists$ to allow marked conclusions as follows:

$$(\vee E): \quad \cfrac{A \vee B \quad \begin{array}{c}\Gamma_1, [A]^i \\ \Pi_1 \\ \langle M\rangle\end{array} \quad \begin{array}{c}\Gamma_2, [B]^i \\ \Pi_2 \\ \langle M\rangle\end{array}}{\langle M\rangle} \; i$$

$$(\exists E): \quad \cfrac{\exists v\, A \quad \begin{array}{c}\Gamma, [A]^i \\ \Pi \\ \langle M\rangle\end{array}}{\langle M\rangle} \; i$$

We thus obtain a natural deduction system ND with FOL rules for formulas of the extended fragment EF. We will extend such a system to handle 'generally'.

3.1 Natural Deduction for Basic 'Generally'

We now consider natural deduction for basic 'generally'.

This extension has three basic rules: elimination and introduction of ∇, as well as a rule transforming marked formulas. Elimination and introduction of ∇ serve to enter and leave a marked environment.

The *elimination* and *introduction* rules for ∇ are given in Table 6. Note that rules (∇E) and $(I\nabla)$ can be used to begin or end a marked track.

---

[10]Note that $\text{rk}(\nabla v\, A(v)) = \text{rk}(A)) + 1$; so $\text{rk}(A)) < \text{rk}(\langle A(\_)\rangle) < \text{rk}(\nabla v\, A(v))$.

[11]For instance, $\langle A(\_)\rangle$ and $\langle B(\_)\rangle$ are sub-formulas of $\langle A(\_) \wedge B(\_)\rangle$ and of $\langle A(\_) \vee B(\_)\rangle$. So, $\langle A(\_)\rangle$ is a sub-formula of $\langle B(\_)\rangle$, when $A(v)$ is a sub-formula of $B(v)$.

[12]Thus, $\langle\top\rangle$ and $\langle\bot\rangle$ are marked formulas.

[13]We consider a system for minimal, intuitionistic or classical logic (Vana, 2008; Vana et al., 2014).

Table 6: Elimination and introduction rules for generalized quantifier ∇

$$(\nabla E) : \frac{\nabla v\, A}{\langle A[v/\_]\rangle} \qquad\qquad (I\nabla) : \frac{\langle A[v/\_]\rangle}{\nabla z\, A[v/z]} \text{ (with } z \notin \operatorname{occ}(A[v/\_])\text{)}$$

Example 3 (Alphabetic variant). *A derivation of* $\nabla z\, r(x,z)$ *from* $\nabla y\, r(x,y)$ *is:*

$$(I\nabla)\ \frac{(\nabla E)\ \dfrac{\nabla y\, r(x,y)}{\langle r(x,\_)\rangle}}{\nabla z\, r(x,z)}$$

Its main track is $\nabla y\, r(x,y)$ (∇E) $\langle r(x,\_)\rangle$ $(I\nabla)$ $\nabla z\, r(x,z)$.

The transformation rule *updown* is given in Table 7 (cf. scheme $[\leftrightarrow \nabla]$).

Table 7: Transformation rule updown

$$
\begin{array}{cc}
\Gamma',[A]^i & \Gamma'',[B]^i \\[4pt]
\Pi' & \Pi''
\end{array}
$$

$$(\updownarrow) : \quad \frac{B \quad \langle A[v/\_]\rangle \quad A}{\langle B[v/\_]\rangle}\ i \qquad\qquad v \notin \operatorname{fr}[\Gamma' \cup \Gamma'']$$

In $(\updownarrow)$, the marked formula $\langle A[v/\_]\rangle$ is its *major premise* and the unmarked formulas A and B are its *minor premises*, while the marked formula $\langle B[v/\_]\rangle$ is its *conclusion*.[14] The track with the marked premise and conclusion is called *central* and derivations Π' and Π'' are called *lateral*. The lateral derivations have higher order than the central track. We often display rule $(\updownarrow)$ as shown in Fig. 3.

Figure 3: Transformation rule updown in abbreviated form

$$[A]^i$$

$$\updownarrow$$

$$(\updownarrow) : \quad \frac{\langle A[v/\_]\rangle \quad [B]^i}{\langle B[v/\_]\rangle}\ i$$

Example 4 (Equivalence scheme). *A derivation of* $\nabla v\, B(v)$ *from* $\forall v\, (A \leftrightarrow B)$ *and* $\nabla v\, A(v)$ *is as follows:*

---

[14]One may regard $(\updownarrow)$ as eliminating the major premise and introducing the conclusion.

Its central track is $\nabla v\, A(v)$ (∇E) $\langle A(\_)\rangle$ $(\updownarrow)$ $\langle B(\_)\rangle$ $(I\nabla)$ $\nabla v\, B(v)$, *with order 0. The lateral track with formulas* $\forall v\,(A \leftrightarrow B)$ *and* $B(v)$ *has order 1, as has the lateral track with formulas* $\forall v\,(A \leftrightarrow B)$ *and* $A(v)$. *See Fig. 4*

Figure 4: Tracks in derivation Π (Example 4: Equivalence scheme)

Tracks	order
$\nabla v\, A(v), \langle A(\_)\rangle, \langle B(\_)\rangle, \nabla v\, B(v)$	0
$\forall v\,(A \leftrightarrow B), B(v)$ & $\forall v\,(A \leftrightarrow B), A(v)$	1
$A(v)$ & $B(v)$	2

The *basic rules* $\mathrm{BR} = \{\,(\nabla E), (I\nabla), (\updownarrow)\,\}$ extend a natural deduction system ND for FOL to a system $\mathsf{ND(BL)}$ for basic logic.

Example 5 (Basic derivation). *Let C be the formula $\forall u \forall z\, [A(u,z) \leftrightarrow B(u,z)]$. We wish to construct a $\mathsf{ND(BL)}$ derivation of $\nabla u \nabla z\, B(u,z)$ from C and $\nabla u \nabla z\, A(u,z)$.*

1. *We start with 2 FOL derivations as follows*
   $\Sigma'\!: C, A(u,z) \vdash_{\mathsf{ND}} B(u,z)$ and $\Sigma''\!: C, B(u,z) \vdash_{\mathsf{ND}} A(u,z)$

2. *Then, we obtain the following derivation Π':*

$$
(I\nabla)\ \cfrac{(\updownarrow)\ \cfrac{\begin{array}{c} C,[A(u,z)]^1 \\ \Sigma' \\ B(u,z) \end{array}}{\langle B(u,\_)\rangle}\quad (\nabla E)\ \cfrac{\nabla z\, A(u,z)}{\langle A(u,\_)\rangle}\quad \cfrac{\begin{array}{c} C,[B(u,z)]^1 \\ \Sigma'' \\ A(u,z) \end{array}}{}\ {}^1}{\nabla z\, B(u,z)}
$$

3. *Similarly, we obtain derivation $\Pi''\!: C, \nabla z\, B(u,z) \vdash_{\mathsf{ND(BL)}} \nabla z\, A(u,z)$.*

4. *From Π' and Π'', we obtain derivation $\Pi\!: C, \nabla u \nabla z\, A(u,z) \vdash_{\mathsf{ND(BL)}}$*
   $\nabla u \nabla z\, B(u,z)$

$$
(I\nabla)\ \cfrac{(\updownarrow)\ \cfrac{\begin{array}{c} C,[\nabla z\, A(u,z)]^3 \\ \Pi' \\ \nabla z\, B(u,z) \end{array}}{\langle \nabla z\, B(\_,z)\rangle}\quad (\nabla E)\ \cfrac{\nabla u \nabla z\, A(u,z)}{\langle \nabla z\, A(\_,z)\rangle}\quad \cfrac{\begin{array}{c} C,[\nabla z\, B(u,z)]^3 \\ \Pi'' \\ \nabla z\, A(u,z) \end{array}}{}\ {}^3}{\nabla u \nabla z\, B(u,z)}
$$

The *structure of the $\mathsf{ND(BL)}$ derivation Π is shown in Fig. 5.*
Its central track is $\nabla u \nabla z\, A$ (∇E) $\nabla z\, A(\_,z)$ $(\updownarrow)$ $\nabla z\, B(\_,z)$ $(I\nabla)$ $\nabla u \nabla z\, B$, with order 0. The track with formulas $\nabla z\, A, \langle A(u,\_)\rangle, \langle B(u,\_)\rangle$ and $\nabla z\, B$ has order 1, whereas the two tracks from C and A to B have order 2.

$$
\dfrac{C,[A]^1}{\Sigma'}\quad \dfrac{[\nabla z A]^3}{\langle A(u,\_)\rangle}
$$

$$
\dfrac{C,[B]^1}{\Sigma''}\quad A
$$

$$
\dfrac{C,[B]^2}{\Sigma''}\quad A \qquad \dfrac{[\nabla z B]^3}{\langle B(u,\_)\rangle}
$$

$$
\dfrac{C,[A]^2}{\Sigma'}\quad B
$$

$$
(\oplus)\ \dfrac{\dfrac{B}{\langle B(u,\_)\rangle}}{\nabla z B}
\qquad \dfrac{\dfrac{A}{\nabla u\nabla z A}\quad {}^{1}}{\langle \nabla z A(\_,z)\rangle}
$$

$$
(\oplus)\ \dfrac{\langle A(u,\_)\rangle}{\nabla z A}\ {}^{2}
$$

$$
\dfrac{\langle \nabla z B(\_,z)\rangle}{\nabla u\nabla z B}\ {}^{3}
$$

Figure 5: Structure of basic derivation II (Example 5)

3.2 Natural Deduction for Specific 'Generally'

We now construct natural deduction for specific versions of 'generally'.

For this purpose, we add to ND(BL) proper and operational rules corresponding to the axioms and schemes in 2.2: Logics for 'Generally'.

Table 8 gives the *proper rules* (cf. Table 4: Proper schemes and axioms).

Table 8: Proper rules: marked $\top$ introduction and marked $\bot$ elimination

$$(I\top): \quad \frac{\top}{\langle\top\rangle} \qquad\qquad (\bot E): \quad \frac{\langle\bot\rangle}{\bot}$$

Example 6 (Proper derivations). *We have derivations for the proper schemes.*

1. *From* $\forall v\, A(v)$ *to* $\nabla v\, A(v)$*, we have the proper derivation* Π_1:

$$(I\nabla)\ \cfrac{(\updownarrow)\ \cfrac{[\top]^1 \qquad (\forall E)\ \cfrac{\forall v\, A(v)}{A(v)} \qquad (I\top)\ \cfrac{\top}{\langle\top\rangle} \qquad \cfrac{}{\top} \qquad [A(v)]^1}{\langle A(\_)\rangle}\ 1}{\nabla v\, A(v)}$$

2. *We have a FOL derivation* $\Pi : \neg\exists v\, \neg A(v) \vdash \neg\neg A(v)$*.*[15] *From* Π *we can obtain the proper derivation* Π_2 *from* $\neg\exists v\, \neg A(v)$ *to* $\neg\nabla v\, \neg A(v)$ *as follows:*

$$(I\neg)\ \cfrac{(\bot E)\ \cfrac{(\updownarrow)\ \cfrac{(\neg E)\ \cfrac{\begin{array}{c}\neg\exists v\, \neg A(v)\\ \Pi \\ \neg\neg A(v) \qquad [\neg A(v)]^1\end{array}}{\bot} \qquad (\nabla E)\ \cfrac{[\nabla v\, \neg A]^2}{\langle \neg A(\_)\rangle} \qquad (I\neg)\ \cfrac{[\bot]^1}{\neg A(v)}}{\langle\bot\rangle}\ 1}{\bot}}{\neg\nabla v\, \neg A(v)}\ 2$$

The central tracks in these derivations go from $\top$ *to* $\nabla v\, A(v)$*, in* Π_1*, and from* $\nabla v\, \neg A$ *to* $\neg\nabla v\, \neg A(v)$*, in* Π_2. ♮

Table 9 gives the *operational rules* (cf. Table 5: Specific schemes).

---

[15]Derivation $\Pi : \neg\exists v\, \neg A(v) \vdash \neg\neg A(v)$ is as follows:

$$(I\neg)\ \cfrac{(\neg E)\ \cfrac{\neg\exists v\, \neg A(v) \qquad (I\exists)\ \cfrac{[\neg A(v)]^1}{\exists v\, \neg A(v)}}{\bot}}{\neg\neg A(v)}\ 1$$

Table 9: Operational rules

Introduction

$\wedge$ $\qquad$ $(\text{I}\underline{\wedge})$: $\dfrac{\langle M \rangle\ \langle N \rangle}{\langle M \wedge N \rangle}$

$\vee$ $\qquad$ $(\text{I}\underline{\vee})$: $\dfrac{\langle M \rangle\ \langle N \rangle}{\langle M \vee N \rangle}$

$\neg$ $\qquad$ $(\text{I}\underline{\neg})$: $\dfrac{\begin{array}{c}[\langle M \rangle]^i \\ \vdots \\ \bot \end{array}}{\langle \neg M \rangle}\ i$

Elimination

$\wedge$ $\qquad$ $(\underline{\wedge}\text{E})$: $\dfrac{\langle M \wedge N \rangle}{\langle M \rangle}\quad \dfrac{\langle M \wedge N \rangle}{\langle N \rangle}$

$\vee$ $\qquad$ $(\underline{\vee}\text{E})$: $\dfrac{\langle M \vee N \rangle \quad \begin{array}{c}[\langle M \rangle]^i \\ \vdots \\ F \end{array} \quad \begin{array}{c}[\langle N \rangle]^i \\ \vdots \\ F \end{array}}{F}\ i$

$\neg$ $\qquad$ $(\underline{\neg}\text{E})$: $\dfrac{\langle \neg M \rangle \quad \langle M \rangle}{\bot}$

Note that rules (I$\top$) and ($\bot$E) can be used to begin or end a marked track. Also, rule ($\neg$E) ends a marked track. Rules with marked premises and conclusion, like (I$\wedge$), (I$\vee$) and ($\wedge$E), can be applied within a marked track.

These rules can be used to formulate logics like those in 2.2 (cf. Table 5). With rule (I$\wedge$), we can derive ∇v$(A(v) \wedge B(v))$ from $(\nabla$v$A(v) \wedge \nabla$v$B(v))$ by applying ($\wedge$E), (∇E), (I$\wedge$) and (I∇).[16] Thus $\vdash_{\text{ND}((\text{I}\wedge))}$ [$\nabla\wedge$]. Similarly $\vdash_{\text{ND}((\wedge\text{E}))}$ [$\wedge\nabla$], $\vdash_{\text{ND}((\text{I}\vee))}$ [$\nabla\vee$], $\vdash_{\text{ND}((\vee\text{E}))}$ [$\vee\nabla$], $\vdash_{\text{ND}((\text{I}\neg))}$ [$\nabla\neg$] and $\vdash_{\text{ND}((\neg\text{E}))}$ [$\neg\nabla$].[17]

Example 7 (Lattice derivation). *If we use '$L(v)$' for "v has legs shaved" and '$F(v)$' for "v has face shaved', then we can formulate some of the assertions in Example 1 (Notions of 'generally') as follows.*

(α) *"Several Brazilians have their legs shaved" by ∇v$L(v)$.*

(β) *"Several Brazilians have their faces shaved" by ∇v$F(v)$.*

($\sqcap$) *"Several Brazilians shave their legs and their faces" by ∇v$(L(v) \wedge F(v))$.*

A lattice derivation Π of ∇v$(L(v) \wedge F(v))$ from (α) and (β) is as follows:

$$(\text{I}\nabla) \cfrac{(\text{I}\wedge) \cfrac{(\nabla\text{E}) \cfrac{\nabla\text{v}\,L(v)}{\langle L(\_)\rangle} \qquad (\nabla\text{E}) \cfrac{\nabla\text{v}\,F(v)}{\langle F(\_)\rangle}}{\langle L(\_) \wedge F(\_)\rangle}}{\nabla\text{v}\,(L(v) \wedge F(v))}$$

This derivation Π has 2 main tracks (with order 0): from the hypotheses ∇v$L(v)$ and ∇v$F(v)$ to the conclusion ∇v$(L(v) \wedge F(v))$.

A lattice derivation like Π of Example 7 also shows that ∇v$(L(v) \wedge \neg D(v))$ (for "Most naturals are larger than fifteen and do not divide twelve") is a lattice consequence of ∇v$L(v)$ (for "Most naturals are larger than fifteen") and ∇v$\neg D(v)$ (for "Most naturals do not divide twelve").

We can also formulate directly Superset ($\sqsupseteq$). For this purpose, we use a transformation rule ($\Downarrow$), similar to ($\Updownarrow$). The next example introduces rule ($\Downarrow$).

---

[16]Indeed, we have the derivation:

$$(\text{I}\nabla) \cfrac{(\text{I}\wedge) \cfrac{(\nabla\text{E}) \cfrac{(\wedge\text{E}) \cfrac{\nabla\text{v}\,(A(v) \wedge B(v))}{\nabla\text{v}\,A(v)}}{\langle A(\_)\rangle} \qquad (\nabla\text{E}) \cfrac{(\wedge\text{E}) \cfrac{\nabla\text{v}\,(A(v) \wedge B(v))}{\nabla\text{v}\,B(v)}}{\langle B(\_)\rangle}}{\langle A(\_) \wedge B(\_)\rangle}}{\nabla\text{v}\,(A(v) \wedge B(v))}$$

[17]For instance, we can derive $\neg \nabla$v$A(v)$ from ∇v$\neg A(v)$ by applying ($\neg$E), (∇E), (I$\neg$) and (I∇).

Example 8 (Upclosed derivation). *Consider the argument in Example 2 (Argument with 'many'). By using 'S(v)' for "v likes sports" and 'T(v)' for "v watches channel SportTV", we can formulate its three assertions as follows:*

$$(\upsilon)\colon \forall v\,(S(v) \to T(v)) \qquad (\sigma)\colon \nabla v\,S(v) \qquad (\tau)\colon \nabla v\,T(v)$$

A derivation of $\nabla v\,T(v)$ from $\forall v\,(S(v) \to T(v)\,)$ and $\nabla v\,S(v)$ is as follows:

$$(\mathrm{I}\nabla)\ \cfrac{(\Downarrow)\ \cfrac{(\nabla\mathrm{E})\ \cfrac{\nabla v\,S(v)}{\langle S(\_)\rangle} \qquad (\to\mathrm{E})\ \cfrac{[\,S(v)\,]^1 \qquad (\forall\mathrm{E})\ \cfrac{\forall v\,(S(v) \to T(v))}{S(v) \to T(v)}}{T(v)}\ 1}{\langle T(\_)\rangle}}{\nabla v\,T(v)}$$

Its main track is $\nabla v\,S(v)\ (\nabla\mathrm{E})\ \langle S(\_)\rangle\ (\Downarrow)\ \langle T(\_)\rangle\ (\mathrm{I}\nabla)\ \nabla v\,T(v)$.

We also have an upclosed derivation from $\nabla v\,L(v)$ to $\nabla v\,(L(v) \vee M(v))$ (cf. Example 1: Notions of 'generally').

The transformation rule *down* is given in Table 10.

Table 10: Transformation rule down

$$\Gamma, [\,A\,]^i$$

$$\Pi$$

$$(\Downarrow)\colon \quad \cfrac{\langle A[v/\_]\rangle \qquad B}{\langle B[v/\_]\rangle}\ i \qquad\qquad v \notin \mathrm{fr}[\Gamma]$$

In ($\Downarrow$), the marked formula $\langle A[v/\_]\rangle$ is its *major premise* and the unmarked formula B is its *minor premise*, while the marked formula $\langle B[v/\_]\rangle$ is its *conclusion*.[18] The track with the marked premise and conclusion is called *central* and derivation Π is called *lateral*. The lateral derivation has higher order than the central track. We often display rule ($\Downarrow$) as shown in Fig. 6.

Figure 6: Transformation rule down in abbreviated form

$$[\,A\,]^i$$

$$\Downarrow$$

$$(\Downarrow)\colon \quad \cfrac{\langle A[v/\_]\rangle \qquad B}{\langle B[v/\_]\rangle}\ i$$

---

[18]One may regard ($\Downarrow$) as eliminating the major premise and introducing the conclusion.

Example 9 (Filter derivation). *Consider the following formulas.*
σ_1: $\nabla v\, S_1(v)$, σ_2: $\nabla v\, S_2(v)$ *and* τ: $\nabla v\, T(v)$ *(sets S_1, S_2 and T in filter $\mathcal{F}$).*
v: $\forall v\,[(S_1(v) \wedge S_2(v)) \to T(v)]$ *(set inclusion $S_1 \cap S_2 \subseteq T$).*
We can derive $\langle T(\_) \rangle$ *from* $\langle S_1(\_) \rangle$, $\langle S_2(\_) \rangle$ *and* $\forall v\,[(S_1(v) \wedge S_2(v)) \to T(v)]$ *by the following derivation* Π:

Notice that the tracks from $\langle S_1(\_)\rangle$ *and* $\langle S_2(\_)\rangle$ *to* $\langle T(\_)\rangle$ *have*
$(I\triangle)$; $(\Downarrow)$.
From Π, *we can obtain the following derivation* Π':

$$(\nabla E)\ \frac{\nabla v\, S_1(v)}{\langle S_1(\_)\rangle} \qquad (\nabla E)\ \frac{\nabla v\, S_2(v)}{\langle S_2(\_)\rangle}$$

$$\Pi$$

$$(I\nabla)\ \frac{\langle T(\_)\rangle}{\nabla v\, T(v)}$$

We thus have a filter derivation of $\nabla v\, T(v)$ *from the formulas* $\nabla v\, S_1(v)$
, $\nabla v\, S_2(v)$ *and* $\forall v\,[(S_1(v) \wedge S_2(v)) \rightarrow T(v)]$. *Its structure appears in*
Fig. 7.

Figure 7: Structure of filter derivation Π' (Example 9)

$$(\Downarrow)\ \frac{\dfrac{\dfrac{\nabla v\, S_1(v)}{\langle S_1(\_)\rangle}\quad \dfrac{\nabla v\, S_2(v)}{\langle S_2(\_)\rangle}}{\dfrac{\langle S_1(\_) \wedge S_2(\_)\rangle}{\langle T(\_)\rangle}}\qquad \dfrac{[\,S_1(v) \wedge S_2(v)\,]^1 \quad \dfrac{\upsilon}{(S_1(v) \wedge S_2(v)) \rightarrow T(v)}}{T(v)}^{1}}{\nabla v\, T(v)}$$

Track	order
$\nabla v\, S_i(v), \langle S_i(\_)\rangle, \langle S_1(\_) \wedge S_2(\_)\rangle, \langle T(\_)\rangle, \nabla v\, T(v)$	0
$\upsilon, (S_1(v) \wedge S_2(v)) \rightarrow T(v), T(v)$	1
$S_1(v) \wedge S_2(v)$	2

The transformation rules for 'generally' are summarized in Table 11.

Table 11: Transformation rules for 'generally'

$$(\Updownarrow):\ \frac{\langle A[v/\_]\rangle \qquad [B]^i}{\langle B[v/\_]\rangle}\,i \qquad\qquad (\Downarrow):\ \frac{\langle A[v/\_]\rangle \qquad B}{\langle B[v/\_]\rangle}\,i$$

In the sequel, we will examine the structure of natural deduction derivations.

4 Natural Deduction for 'Generally': Analysis

We now analyze the inner structure of natural deduction derivations.

We will examine reductions (in 4.1), analyze systems for some families (in 4.2) and then specialize these results to some specific cases (in 4.3).

We can formulate natural deduction systems for various logics of 'generally'. For this purpose, we first consider the following set Ω of rules:

$$\Omega \;\coloneqq\; \underset{\nabla}{(\nabla E);(I\nabla)} \quad \underset{\text{proper}}{(I\underline{\top});(\underline{\bot}E)} \quad \underset{\text{operational}}{(I\wedge),(\wedge E),(I\vee),(\vee E),(I\neg),(\neg E)}$$

We then use a subset $\Xi \subseteq \Omega$ and a transformation rule $(\underline{\text{Tr}}) \in \{\,(\updownarrow),(\Downarrow)\,\}$ to form a system $\Xi_{\underline{\text{Tr}}} \coloneqq \mathsf{ND}(\Xi \cup \{(\underline{\text{Tr}})\})$.

We will use the notation '.*' for "a finite (maybe 0) number of applications".

Derivations in such systems may use rules from the following sets.

(Elm) *Elimination rules*, consisting of the elimination rules of ND together with $\underline{\text{Elm}} \coloneqq \{\,(\nabla E),(\underline{\bot}E),(\wedge E),(\vee E),(\neg E)\,\}$.

(Int) *Introduction rules*, consisting of the introduction rules of ND together with $\underline{\text{Int}} \coloneqq \{\,(I\nabla),(I\underline{\top}),(I\wedge),(I\vee),(I\neg)\,\}$.

(Tr) *Transformation rules* from $\underline{\text{Tr}} \coloneqq \{\,(\updownarrow),(\Downarrow)\,\}$.

Such derivations may have special segments as follows (see Fig. 8).

($\searrow$) *Descending* segments, consisting of consecutive applications of elimination rules: $H\,(\text{Elm})^*\,G$. Note that final G is a sub-formula of initial H.

($\nearrow$) *Ascending* segments, consisting of consecutive applications of elimination rules: $G\,(\text{Int})^*\,F$. Note that initial G is a sub-formula of final F.

($\dashrightarrow$) *Flat* segments, consisting of a single application of a transformation rule: $\langle M \rangle\,(\underline{\text{Tr}})\,\langle N \rangle$.

Figure 8: Special segments

Descending   Flat   Ascending

H F

$\quad\searrow$

$\quad\quad G$ $\langle M \rangle \;\dashrightarrow\; \langle N \rangle$ $\quad\nearrow$

$\quad\quad\quad\quad\quad\quad\quad\quad\quad\quad\quad\quad\quad\quad\quad\quad\quad\quad\quad\quad\quad\quad\quad\quad G$

4.1 Natural Deduction for 'Generally': Reductions

We now examine reductions for natural deduction derivations: involving ∇-rules, transformation rules and operational rules.

Lemma 1 (Reduction for ∇). *We have the following reduction for ∇ (see Fig. 9):*

$[\text{I}\nabla\,;\,\nabla\text{E})$ *Introduction followed by elimination of ∇: contract* $(\text{I}\nabla)\,;\,(\nabla\text{E})$

So, we can contract the cut segment $\langle A(\_)\rangle\,(\text{I}\nabla)\,\nabla z\,A(z)\,(\nabla\text{E})\,\langle A(\_)\rangle$ *to* $\langle A(\_)\rangle$. ♭

Figure 9: Reduction $[\text{I}\nabla\,;\,\nabla\text{E})$ for ∇

$$
(\nabla\text{E})\ \frac{(\text{I}\nabla)\ \dfrac{\overset{\Pi}{\langle A(\_)\rangle}}{\nabla z\,A(z)}}{\langle A(\_)\rangle}\qquad\Rrightarrow\qquad \overset{\Pi}{\langle A(\_)\rangle}
$$

Lemma 2 (Transformation reductions). *We have the* ($\underline{\text{Tr}}$) *reductions.*

$[\updownarrow\,;\,\updownarrow)$ *Consecutive applications of updown rule:* $(\updownarrow)\,;\,(\updownarrow)\Rrightarrow(\updownarrow)$.
 Thus: $\langle M\rangle\,(\updownarrow)\,\langle N\rangle\,(\updownarrow)\,P\Rrightarrow\langle M\rangle\,(\updownarrow)\,\langle P\rangle$.

$[\Downarrow\,;\,\Downarrow)$ *Consecutive applications of down rule:* $(\Downarrow)\,;\,(\Downarrow)\Rrightarrow(\Downarrow)$.
 Thus: $\langle M\rangle\,(\Downarrow)\,\langle N\rangle\,(\Downarrow)\,\langle P\rangle\Rrightarrow\langle M\rangle\,(\Downarrow)\,\langle P\rangle$.

These reductions involve derivation gluing and may introduce new cut formulas, but only in lateral tracks (with higher order). ♭

Proof. See Figs. 10 and 11.

□

Figure 10: Transformation reduction [⇕ ; ⇕] for consecutive applications of (⇕) rule

$$
(\Uparrow)\;\;\cfrac{
  \cfrac{[B(y)]^j}{\begin{array}{c}\Pi_3\\ C(y)\end{array}}\quad
  (\Uparrow)\;\cfrac{
    \cfrac{[A(x)]^i}{\begin{array}{c}\Pi_1\\ B(x)\end{array}}\;\langle A(\_)\rangle \qquad
    \cfrac{[B(x)]^i}{\begin{array}{c}\Pi_2\\ A(x)\end{array}}
  }{\langle B(\_)\rangle}\;i
}{\langle C(\_)\rangle}\;j
$$

$$\Longrightarrow$$

$$
(\Uparrow)\;\;\cfrac{
  \cfrac{[C(y)]^j}{\begin{array}{c}\Pi_4\\ B(y)\end{array}}\quad
  (\Uparrow)\;\cfrac{
    \cfrac{[A(z)]^k}{\begin{array}{c}\Pi_1'\\ B(z)\\ \Pi_3'\\ C(z)\end{array}}\;\langle A(\_)\rangle \qquad
    \cfrac{[C(z)]^k}{\begin{array}{c}\Pi_4'\\ B(z)\\ \Pi_2'\\ F(z)\end{array}}
  }{\langle C(\_)\rangle}
}{}\;
$$

Here, $\Pi_1' := \Pi_1[x/z]$, $\Pi_2' := \Pi_2[x/z]$, $\Pi_3' := \Pi_3[y/z]$, $\Pi_4' := \Pi_4[y/z]$, where z is a new variable.

Figure 11: Transformation reduction $[\Downarrow ; \Downarrow]$ for consecutive applications of $(\Downarrow)$ rule

$$
(\Downarrow)\cfrac{\langle A(\_)\rangle \quad \cfrac{[A(x)]^i}{\begin{array}{c}\Pi_1\\ B(x)\end{array}}}{\cfrac{\langle B(\_)\rangle}{\quad}\ i \qquad \cfrac{[B(y)]^j}{\begin{array}{c}\Pi_2\\ C(y)\end{array}}}{\langle C(\_)\rangle}\ j
$$

$$
\Rrightarrow
$$

$$
(\Downarrow)\ \cfrac{\langle A(\_)\rangle \qquad \begin{array}{c}[A(z)]^k\\ \Pi_1'\\ B(z)\\ \Pi_2'\\ C(z)\end{array}}{\langle C(\_)\rangle}\ k
$$

Here, $\Pi_1' := \Pi_1[x/z]$ and $\Pi_2' := \Pi_2[y/z]$, where z is a new variable.

Remark 1 (Operational reductions). *We have the following three operational reductions for tracks with marked introduction followed by marked elimination.*

$[\underline{\wedge}]$ $\langle M_i \rangle (\text{I}\underline{\wedge}) \langle M_1 \wedge M_2 \rangle (\underline{\wedge}\text{E}) \langle M_i \rangle \Rrightarrow \langle M_i \rangle$ *(see Fig. 12)*.

$[\underline{\vee}]$ $\langle M_i \rangle (\text{I}\underline{\vee}) \langle M_1 \vee M_2 \rangle (\underline{\vee}\text{E}) F \Rrightarrow \langle M_i \rangle \ldots F$ *(see Fig. 13)*.

$[\underline{\neg}]$ $\langle M \rangle (\text{I}\underline{\neg}) \langle \neg M \rangle (\underline{\neg}\text{E}) \bot \Rrightarrow \langle M \rangle \ldots \bot$ *(see Fig. 14)*.

These reductions are similar to those for FOL (Vana, 2008; Vana et al., 2007, 2014).

Figure 12: Operational reduction $[\underline{\wedge}]$ for marked $\wedge$

$$(\underline{\wedge}\text{E}) \ \dfrac{(\text{I}\underline{\wedge}) \ \dfrac{\overset{\Pi_1}{\langle M_1 \rangle} \quad \overset{\Pi_2}{\langle M_2 \rangle}}{\langle M_1 \wedge M_2 \rangle}}{\langle M_i \rangle} \qquad \Rrightarrow \qquad \overset{\Pi_i}{\langle M_i \rangle}$$

Figure 13: Operational reduction $[\underline{\vee}]$ for marked $\vee$

$$(\underline{\vee}\text{E}) \ \dfrac{(\text{I}\underline{\vee}) \dfrac{\overset{\Pi_1}{\langle M_1 \rangle} \quad \overset{\Pi_2}{\langle M_2 \rangle}}{\langle M_1 \vee M_2 \rangle} \qquad \overset{[\langle M_1 \rangle]^i}{\overset{\Pi'_1}{F}} \qquad \overset{[\langle M_2 \rangle]^i}{\overset{\Pi'_2}{F}}}{F} \, i \qquad \Rrightarrow \qquad \overset{\Pi_i}{\overset{\langle M_i \rangle}{\overset{\Pi'_i}{F}}}$$

Figure 14: Operational reduction $[\underline{\neg}]$ for marked $\neg$

$$
(\neg E) \quad \cfrac{(I_{\neg}) \quad \cfrac{\begin{array}{c}[\langle M \rangle]^i \\ \Pi' \\ \bot\end{array}}{\langle \neg M \rangle} \, i \qquad \begin{array}{c}\Pi'' \\ \langle M \rangle\end{array}}{\bot} \qquad\Longrightarrow\qquad \begin{array}{c}\Pi'' \\ \langle M \rangle \\ \Pi' \\ \bot\end{array}
$$

We will employ the notation '$.^{\leq 1}$' for "at most one application".

Corollary 1 (Iterated reductions). *We can reduce applications of the rules as follows:* $\{(\text{Bm}), (\underline{\text{Tr}}), (\text{ht})\}^* \Longrightarrow (\text{Bm})^* \, ; \, [(\underline{\text{Int}})^* \, ; \, (\underline{\text{Tr}})^{\leq 1} \, ; \, (\underline{\text{Elm}})^*]^* \, ; \, (\text{ht})^*.$

Proof. By the reduction results: Lemmas 1 and 2 and Remark 1. $\qquad\square$

Now, consider a natural deduction system $\text{ND}(\Omega \cup \{(\underline{\text{Tr}})\})$ as above. Much as in FOL, such systems are normalizable. But, normal derivations may have several (non-consecutive) applications of a transformation rule, with the aspect shown in Section 1 (Introduction). So, we cannot guarantee the sub-formula property.

Example 10 (Derivation with 2 updowns). *Given unary predicate symbols p_i, q_i and s_i, for $i = 1, 2$, and t, consider the following formulas:*

α: $\forall v \, [(p_1(v) \land p_2(v)) \leftrightarrow (q_1(v) \lor q_2(v))],$
β_i: $\forall v \, [q_i(v) \to s_1(v)]$ *(for $i = 1, 2$),*
γ: $\forall v \, [(s_1(v) \land s_2(v)) \leftrightarrow t(v)].$

We have a derivation Π of $\nabla v \, t(v)$ from $\Delta = \{\alpha, \beta_1, \beta_2, \gamma, \nabla v \, p_1(v), \nabla v \, p_2(v)\}$. This derivation Π can be constructed as follows.

1. *We construct derivation Π_1 from $\Delta_1 = \{\alpha, \nabla v \, p_1(v), \nabla v \, p_2(v)\}$ to $\langle q_1(\_) \land q_2(\_)\rangle$:*

$$
(\updownarrow) \quad \cfrac{(I_\wedge) \quad \cfrac{(\nabla E) \, \cfrac{\nabla v \, p_1(v)}{\langle p_1(\_)\rangle} \qquad (\nabla E) \, \cfrac{\nabla v \, p_2(v)}{\langle p_2(\_)\rangle}}{\langle p_1(\_) \land p_2(\_)\rangle}}{\langle q_1(\_) \lor q_2(\_)\rangle}
$$

$$
\begin{array}{c}[p_1(v) \land p_2(v)]^1 \qquad \alpha \\[4pt] \begin{array}{c}\wedge \\ \vdots \\ \vee\end{array} \\[4pt] \cfrac{[q_1(v) \lor q_2(v)]^1}{} 1\end{array}
$$

2. *Finally, from Π_2, we construct derivation Π from $\Delta = \Delta_2 \cup \{\gamma\}$ to $\nabla v \, t(v)$:*

From Π_1, we construct derivation Π_2 from $\Delta_2 := \Delta_1 \cup \{\beta_1, \beta_2\}$ to $\langle s_1(\_) \wedge s_2(\_)\rangle$:

$$
\cfrac{
\cfrac{
\cfrac{\Pi_1}{\langle q_1(\_) \vee q_2(\_)\rangle}
\qquad
\cfrac{[\langle q_1(\_)\rangle]^2 \quad \beta_1 \;\cdots}{\langle s_1(\_)\rangle}
\qquad
\cfrac{[\langle q_2(\_)\rangle]^2 \quad \beta_2 \;\cdots}{\langle s_1(\_)\rangle}
}{\langle s_1(\_)\rangle}\;(\vee E)^2
\qquad\qquad
\cfrac{\nabla \vee s_2(v)}{\langle s_2(\_)\rangle}\;(\nabla E)
}{\langle s_1(\_) \wedge s_2(\_)\rangle}\;(I\wedge)
$$

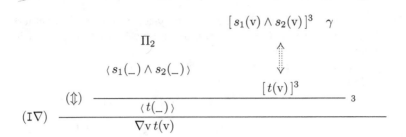

$$[\,s_1(\mathrm{v}) \wedge s_2(\mathrm{v})\,]^3 \quad \gamma$$

$$\Pi_2$$

$$\langle\, s_1(\_) \wedge s_2(\_)\,\rangle$$

$$\begin{array}{c} [\,t(\mathrm{v})\,]^3 \\ \hline \end{array}$$

$$(\updownarrow)\ \dfrac{}{\langle\, t(\_)\,\rangle}\ \ 3$$

$$(\mathrm{I}\nabla)\ \dfrac{}{\nabla \mathrm{v}\, t(\mathrm{v})}$$

Fig. 15 shows the structure of this derivation Π. Its main tracks from $\nabla \mathrm{v}\, p_i(\mathrm{v})$ to $\nabla \mathrm{v}\, t(\mathrm{v})$ have 2 applications of $(\updownarrow)$.

Figure 15: Structure of derivation Π (Example 10: Derivation with 2 up-downs)

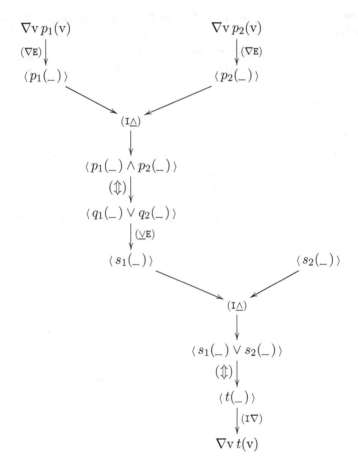

4.2 Natural Deduction for 'Generally': Structure

We now examine natural deduction systems for some specific families. A motivation for this is the usual way to show that a set is in a filter (cf. Section 1).

We will consider systems as follows. We start with the *restricted rule set*:

$$\text{RR} \;\;:=\;\; \begin{array}{ccc} \nabla\text{E}; \text{I}\nabla & \text{I}\underline{\text{T}}; \underline{\bot}\text{E}, & \text{I}\wedge, \triangle\text{E}, \text{I}\underline{\vee} \\[4pt] \nabla & \text{proper} & \text{operational} \end{array}$$

We then form the sets $\text{MR}_{\updownarrow} := \text{RR} \cup \{(\updownarrow)\}$ and $\text{AR}_{\downarrow} := \text{RR} \cup \{(\Downarrow)\}$. The *restricted* natural deduction systems are of the form $\text{ND}(\Xi)$, with $\Xi \subseteq \text{MR}_{\updownarrow}$ or $\Xi \subseteq \text{AR}_{\downarrow}$.

We will use '•' for a binary connective $\bullet \in \{\wedge, \vee\}$.

We first examine some reductions, which rely on simple properties of the binary connectives $\wedge$ and $\vee$.

Lemma 3 (Restricted reductions). *We have the following restricted reductions.*

[$\underline{\text{Tr}}$; I•) *Transformation followed by marked introduction:* ($\underline{\text{Tr}}$) ; (I•) $\Rightarrow$
$\Rightarrow$ (I•) ; ($\underline{\text{Tr}}$), *so:* $\langle M \rangle (\underline{\text{Tr}}) \langle N \rangle (\text{I•}) \langle N \bullet P \rangle \Rrightarrow \langle M \rangle (\text{I•}) \langle M \bullet P \rangle (\underline{\text{Tr}}) \langle N \bullet P \rangle$.

[$\triangle$E ; $\underline{\text{Tr}}$) *Marked $\wedge$ elimination followed by transformation:* ($\triangle$E) ; ($\underline{\text{Tr}}$) $\Rightarrow$
$\Rightarrow$ ($\underline{\text{Tr}}$) ; ($\triangle$E), *so:* $\langle M \wedge N \rangle (\triangle\text{E}) \langle N \rangle (\underline{\text{Tr}}) \langle P \rangle \Rrightarrow \langle M \wedge N \rangle (\underline{\text{Tr}}) \langle M \wedge P \rangle (\triangle\text{E}) \langle P \rangle$.

[$\triangle$E ; I•) *Marked elimination and introduction:* ($\triangle$E) ; (I•) $\Rrightarrow$ (I•) ; ($\underline{\text{Tr}}$) ; ($\triangle$E),
so: $\langle M_1 \wedge M_2 \rangle (\triangle\text{E}) \langle M_i \rangle (\text{I•}) \langle M_i \bullet P \rangle \Rrightarrow$
$\langle M_1 \wedge M_2 \rangle (\text{I•}) \langle (M_1 \wedge M_2) \bullet P \rangle (\underline{\text{Tr}}) \langle (M_1 \bullet P) \wedge (M_2 \bullet P) \rangle (\triangle\text{E}) \langle M_i \bullet P \rangle$.

[$\triangle$E ; $\triangle$E) *Consecutive marked $\wedge$ eliminations:* ($\triangle$E) ; ($\triangle$E) $\Rrightarrow$ ($\underline{\text{Tr}}$) ; ($\triangle$E),
so:
$\langle M \wedge (N \wedge P) \rangle (\triangle\text{E}) \langle N \wedge P \rangle (\triangle\text{E}) \langle P \rangle \Rrightarrow \langle M \wedge (N \wedge P) \rangle (\underline{\text{Tr}}) \langle (M \wedge N) \wedge P \rangle (\triangle\text{E}) \langle P \rangle$.

These reductions may introduce new cut formulas, but only in lateral tracks (with higher order).   ♭

Proof. See Figs. 16, 17, 18 and 19 for ($\Downarrow$). The cases with ($\updownarrow$) are similar.

□

Figure 16: Reduction for down and marked introduction

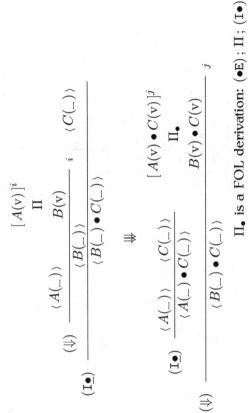

$\Pi_\bullet$ is a FOL derivation: ($\bullet$E) ; Π ; (I$\bullet$)

Figure 17: Reduction for marked ∧ elimination and down

$$
(\Downarrow) \quad \cfrac{(\triangle E)\ \cfrac{\langle M \wedge N \rangle}{\langle N \rangle} \qquad \begin{array}{c} [N]^i \\ \Pi \\ P \end{array}}{\langle P \rangle}\ i
$$

$$\Downarrow$$

$$
(\triangle E)\ \cfrac{(\Downarrow)\ \cfrac{\langle M \wedge N \rangle \quad (\text{I}\wedge)\ \cfrac{(\wedge E)\ \cfrac{[M \wedge N]^j}{M} \qquad (\wedge E)\ \cfrac{\begin{array}{c}[M \wedge N]^j\\ N\end{array}}{\begin{array}{c}\Pi\\ P\end{array}}}{M \wedge P}\ j}{\langle M \wedge P \rangle}}{\langle P \rangle}
$$

Figure 18: Reduction for $(\underline{\wedge}\mathtt{E})$ and $(\mathtt{I}\underline{\bullet})$

$$(\mathtt{I}\underline{\bullet})\ \dfrac{(\underline{\wedge}\mathtt{E})\ \dfrac{\langle M_1 \wedge M_2 \rangle}{\langle M_i \rangle} \qquad \langle P \rangle}{\langle M_i \bullet P \rangle}$$

$$\Downarrow$$

$$(\underline{\wedge}\mathtt{E})\ \dfrac{(\Downarrow)\ \dfrac{(\mathtt{I}\underline{\bullet})\ \dfrac{\langle M_1 \wedge M_2 \rangle \quad \langle P \rangle}{\langle (M_1 \wedge M_2) \bullet P \rangle} \qquad \begin{matrix}[\,(M_1 \wedge M_2) \bullet P\,]^i \\ \Pi \\ (M_1 \bullet P) \wedge (M_2 \bullet P)\end{matrix}}{\langle (M_1 \bullet P) \wedge (M_2 \bullet P) \rangle}\, i}{\langle M_i \bullet P \rangle}$$

Π is a normal FOL derivation: $(\bullet\mathtt{E})$; $(\wedge\mathtt{E})$; $(\mathtt{I}\bullet)$; $(\mathtt{I}\wedge)$

Figure 19: Reduction for consecutive $(\underline{\wedge}\mathtt{E})$'s

$$(\underline{\wedge}\mathtt{E})\ \dfrac{(\underline{\wedge}\mathtt{E})\ \dfrac{\langle M \wedge (N \wedge P) \rangle}{\langle N \wedge P \rangle}}{\langle P \rangle}$$

$$\Downarrow$$

$$(\underline{\wedge}\mathtt{E})\ \dfrac{(\Downarrow)\ \dfrac{\langle M \wedge (N \wedge P) \rangle \qquad \begin{matrix}[\,M \wedge (N \wedge P)\,]^i \\ \Pi \\ (M \wedge N) \wedge P\end{matrix}}{\langle (M \wedge N) \wedge P \rangle}\, i}{\langle P \rangle}$$

Π is a normal FOL derivation: $(\wedge\mathtt{E})^2$; $(\mathtt{I}\wedge)^2$

Example 11 (Reducing multiple updowns). *Consider the following derivation Π:*

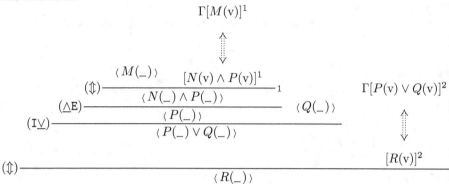

$$\Gamma[M(\mathrm{v})]^1$$

Derivation Π has the form $(\updownarrow)$; $(\triangle E)$; $(I\underline{\vee})$; $(\updownarrow)$. *By* $[\triangle E\,;\,I\underline{\vee})$, $[\updownarrow\,;\,I\triangle)$ *and* $[\updownarrow\,;\,\updownarrow)$, *it can be reduced to a derivation Π' of the form* $(I\underline{\vee})$; $(\updownarrow)$; $(\triangle E)$.
We can obtain Π' from Π as follows:

$(\updownarrow)$; $(\triangle E)$; $(I\underline{\vee})$; $(\updownarrow)$ $\overset{[\triangle E\,;\,I\underline{\vee})}{\Longrightarrow}$ $(\updownarrow)$; $(I\underline{\vee})$; $(\updownarrow)$; $(\triangle E)$; $(\updownarrow)$

$\qquad\qquad\qquad\qquad\qquad [\updownarrow\,;\,I\underline{\vee})\Downarrow \qquad\qquad\qquad\qquad \Downarrow [\triangle E\,;\,\updownarrow)$

$\qquad\qquad\qquad\qquad (I\underline{\vee})$; $(\updownarrow)$; $(\updownarrow)$; $(\updownarrow)$; $(\triangle E)$

$\qquad\qquad\qquad\qquad\qquad\qquad\qquad \Downarrow [\updownarrow\,;\,\updownarrow)$

$\qquad\qquad\qquad\qquad\qquad (I\underline{\vee})$; $(\updownarrow)$; $(\triangle E)$

Derivation Π' is as follows:

$$\Gamma\ [M(\mathrm{v}) \vee Q(\mathrm{v})]^3$$

In Π', *the track from* $\langle M(\_) \vee Q(\_)\rangle$ *to* $\langle R(\_)\rangle$ *has a single* $(\updownarrow)$.

Corollary 2 (Central transformation). *Restricted rule applications can be reduced as follows:* $\{(\underline{\text{Elm}}), (\underline{\text{Tr}}), (\text{Int})\}^* \Rightarrow^* (\text{Elm})^*$; $\{(\underline{\text{Elm}}), (\underline{\text{Tr}}), (\underline{\text{Int}})\}^*$; $(\text{Int})^*$
We can reduce restricted rule applications to a most one central transformation:

1. $\{(\underline{\text{Elm}}), (\underline{\text{Tr}}), (\underline{\text{Int}})\}^* \Rightarrow^* (\underline{\text{Int}})^*$; $(\underline{\text{Tr}})^{\leq 1}$; $(\underline{\text{Elm}})^{\leq 1}$
2. $\{(\text{Elm}), (\underline{\text{Tr}}), (\text{Int})\}^* \Rightarrow^* (\text{Elm})^*$; $(\underline{\text{Int}})^*$; $(\underline{\text{Tr}})^{\leq 1}$; $(\underline{\text{Elm}})^{\leq 1}$; $(\text{Int})^*$.

Proof. By Corollary 1 (Iterated reductions) and Lemma 3.

$\square$

Example 12 (Restricted derivations). *We consider some restricted derivations.*

(0) *Derivation* Π_0 *is as follows:*

$$
(I\nabla)\ \dfrac{(I\wedge)\ \dfrac{(\nabla E)\ \dfrac{(\forall E)\ \dfrac{\forall x_1\,\nabla y_1\,M_1(x_1,y_1)}{\nabla y_1\,M_1(x_1,y_1)}}{\langle\,M_1(x_1,\_)\,\rangle}\qquad (\nabla E)\ \dfrac{(\forall E)\ \dfrac{\forall x_2\,\nabla y_2\,M_2(x_2,y_2)}{\nabla y_2\,M_2(x_2,y_2)}}{\langle\,M_2(x_2,\_)\,\rangle}}{\langle\,M_1(x_1,\_)\,\wedge\,M_1(x_1,\_)\,\rangle}}{\nabla z\,[M_1(x_1,z)\,\wedge\,M_2(x_2,z)]}
$$

Derivation Π_0 *has no application of transformation rule. Its tracks from* $\forall x_i\,\nabla y_i\,M_1(x_i,y_i)$ *to* $\nabla z\,[M_1(x_1,z)\,\wedge\,M_2(x_2,z)]$ *have the aspect:*

$$\forall x_i \nabla y_i M_i$$
$$\searrow$$
$$\nabla y_i M_i \qquad\qquad \left\langle\ \begin{matrix} M_1(x_1,\_) \\ \wedge \\ M_2(x_2,\_) \end{matrix}\ \right\rangle$$
$$\searrow \qquad\qquad \nearrow$$
$$\langle\ M_i(x_i,\_)\ \rangle$$
$$\searrow$$
$$\nabla z\ \left[\ \begin{matrix} M_1(x_1,z) \\ \wedge \\ M_2(x_2,z) \end{matrix}\ \right]$$

Each one of these tracks have a single minimal formula: $\langle\,M_i(x_i,\_)\,\rangle$.

(<) *Derivation* $\Pi_<$ *is as follows:*

$$\Gamma\quad[\,N(v)\,]^1$$

$$
(I\nabla)\ \dfrac{(I\wedge)\ \dfrac{(\nabla E)\ \dfrac{\nabla v\,M(v)}{\langle\,M(\_)\,\rangle}\qquad (\mathop{\updownarrow})\ \dfrac{(\nabla E)\ \dfrac{\nabla v\,N(v)}{\langle\,N(\_)\,\rangle}}{\langle\,P(\_)\,\rangle}\quad [\,P(v)\,]^1}{\langle\,M(\_)\,\wedge\,P(\_)\,\rangle}}{}}{\nabla v\,[M(v)\,\wedge\,P(v)]}\ {}_1
$$

In derivation $\Pi_<$, *the track from* $\nabla v\,N(v)$ *to conclusion* $\nabla v\,[M(v)\,\wedge\,P(v)]$ *has a single application of transformation rule* ($\mathop{\updownarrow}$). *It has the aspect:*

$$\nabla v\,N(v) \qquad\qquad\qquad\qquad \nabla v\,[M(v)\,\wedge\,P(v)]$$
$$\searrow \qquad\qquad\qquad\qquad\qquad \nearrow$$
$$\langle\,N(\_)\,\rangle\ \ \dashrightarrow\ \ \langle\,M(\_)\,\wedge\,P(\_)\,\rangle$$

This track has 2 minimal formulas: $\langle\,N(\_)\,\rangle$ *and* $\langle\,M(\_)\,\wedge\,P(\_)\,\rangle$.

Example 13 (Reduced restricted derivation). *Consider derivation* Π *as follows:*

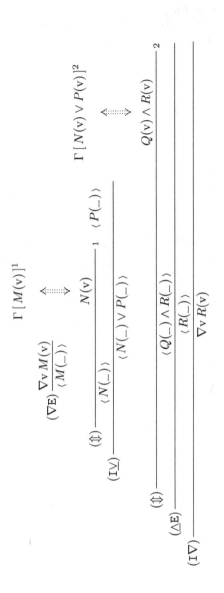

In derivation Π, *the marked segments (from* ⟨ $M(\_)$ ⟩ *and* ⟨ $P(\_)$ ⟩ *to* ⟨ $R(\_)$ ⟩) *have the form* (↕) ; (I∨) ; (↕) ; (∧E). *By* [↕ ; I∨), [∧E ; ↕) *and* [↕ ; ↕), Π *can be reduced to a derivation* Π′ *with marked segments of the form* (I∨) ; (↕) ; (∧E), *namely:*

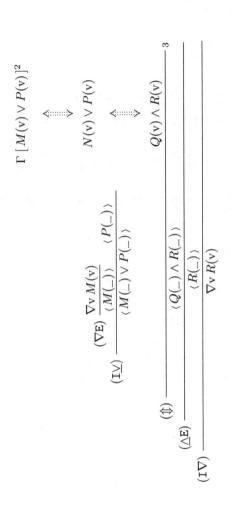

In Π′, the track from hypothesis ∇v M(v) to conclusion ∇v Q(v) has the aspect:

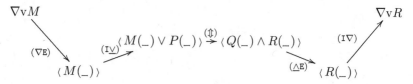

This track has 2 minimal formulas (⟨M(_)⟩ and ⟨R(_)⟩) and 2 large formulas (⟨M(_) ∨ P(_)⟩ and ⟨Q(_) ∧ R(_)⟩).

Proposition 1 (Restricted marked tracks). *We can reduce restricted marked tracks (with rule applications) to one of the following forms.*

(0) *No application of transformation rule:* ⟨M⟩ (Elm)* ⟨N⟩ (Int)* ⟨P⟩

     ⟨M⟩   ⟨P⟩
        ↘   ↗
          ⟨N⟩

(<) *A single (Tr) before introductions:* ⟨M⟩ (Elm)* ⟨N⟩ (Tr) ⟨P⟩ (Int)* ⟨Q⟩

    ⟨M⟩   ⟨Q⟩
      ↘   ↗
       ⟨N⟩ --→ ⟨P⟩

(>) *A single (Tr) after introductions:* ⟨M⟩ (Int)* ⟨N⟩ (Tr) ⟨P⟩ (Elm)$^{≤1}$ ⟨Q⟩

       ⟨N⟩ --→ ⟨P⟩
      ↗   ↘
    ⟨M⟩   ⟨Q⟩

Proof. By Corollary 2 (Central transformation): we can reduce {(Elm), (Tr), (Int)}* to (Elm)*; (Tr)$^{≤1}$; (Int)* or to (Int)*; (Tr)$^{≤1}$; (Elm)$^{≤1}$. □

Theorem 1 (Restricted derivation tracks). *We can reduce restricted derivation tracks (with rule applications) to one of the following forms.*

(0) *No application of transformation rule (1 minimal formula):*

     H   F
       ↘   ↗
        G   *minimal formula:* G

(<) *A single (Tr) before introductions (2 minimal marked formulas):*

minimal formulas: $\langle N \rangle, \langle P \rangle$

(>) *A single* ($\underline{\text{Tr}}$) *after introductions (2 large and 2 minimal formulas):*

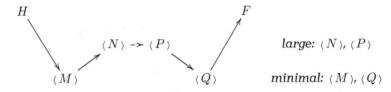

large: $\langle N \rangle, \langle P \rangle$

minimal: $\langle M \rangle, \langle Q \rangle$

Proof. By Corollary 2 (Central transformation) and Proposition 1.

□

4.3 Natural Deduction for 'Generally': Special Cases

We can now specialize the results of our analysis to some special cases.

We will examine basic logic and its extensions to proper modules as well as to (proper) upclosed families, lattices and filters. With the sets NR $= \{(\nabla E), (I\nabla)\}$ and PR $= \{(I\underline{\top}), (\underline{\bot}E)\}$, we will consider the following formulations for these logics.

(BL) Basic logic $\text{BL}_{\updownarrow} := \text{ND}(\text{NR} \cup \{(\updownarrow)\})$.

(PL) Proper logic $\text{PL}_{\updownarrow} := \text{ND}(\text{NR} \cup \text{PR} \cup \{(\updownarrow)\})$.

(SL) Upclosed logic $\text{SL}_{\downarrow} := \text{ND}(\text{NR} \cup \text{PR} \cup \{(\Downarrow)\})$.

(LL) Lattice logic $\text{LL}_{\updownarrow} := \text{ND}(\text{NR} \cup \text{PR} \cup \{(I\underline{\wedge}), (I\underline{\vee})\} \cup \{(\updownarrow)\})$.

(FL) Filter logic $\text{FL}_{\downarrow} := \text{ND}(\text{NR} \cup \text{PR} \cup \{(I\underline{\wedge})\} \cup \{(\Downarrow)\})$.

In each case, we have a *growing* system: a restricted system without ($\wedge$E). Each such system is formulated as $\text{ND}(\Xi \cup \{(\underline{\text{Tr}})\})$, with ($\underline{\text{Tr}}$) $\in \{(\updownarrow), (\Downarrow)\}$ and $\Xi \subseteq \text{NR} \cup \text{PR} \cup \{(I\underline{\wedge})\}$. Let us examine such growing systems.

Example 14 (Reduced growing derivation). *Consider the following derivation* Π:

Why is this a Proof? Festschrift for Luiz Carlos Pereira

152

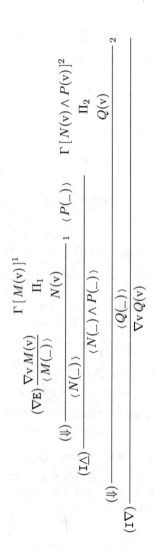

$$(\nabla E)\ \dfrac{\nabla v\, M(v)}{\langle M(\_)\rangle}\quad \dfrac{\Gamma\,[\,M(v)\,]^1}{\Pi_1}\ N(v)$$

$$(\Downarrow)\ \dfrac{}{\langle N(\_)\rangle}\qquad 1\ \langle P(\_)\rangle$$

$$(\triangle I)\ \dfrac{}{\langle N(\_)\rangle \wedge P(\_)\rangle}\qquad \dfrac{\Gamma\,[\,N(v)\wedge P(v)\,]^2}{\Pi_2}\ Q(v)$$

$$(\Uparrow)\ \dfrac{}{\langle Q(\_)\rangle}$$

$$(\triangle I)\ \dfrac{}{\nabla v\, Q(v)}\qquad 2$$

In derivation Π, the marked segments (from $\langle M(\_)\rangle$ and $\langle P(\_)\rangle$ to $\langle Q(\_)\rangle$) have the form $(\Downarrow)\,;\,(\mathrm{I}\wedge)\,;\,(\Downarrow)$. By $[\Downarrow\,;\,\mathrm{I}\wedge]$ and $[\Downarrow\,;\,\Downarrow]$, derivation Π can be reduced to a derivation Π' with marked segments of the form $(\mathrm{I}\wedge)\,;\,(\Downarrow)$, namely:

$$
(\mathrm{I}\nabla)\ \dfrac{(\Downarrow)\ \dfrac{(\mathrm{I}\wedge)\ \dfrac{(\nabla\mathrm{E})\ \dfrac{\nabla\mathrm{v}\,M(\mathrm{v})}{\langle M(\_)\rangle}\quad \langle P(\_)\rangle}{\langle M(\_)\wedge P(\_)\rangle}}{\langle Q(\_)\rangle}\qquad \dfrac{\Gamma\ [M(\mathrm{v})\wedge P(\mathrm{v})]^2\quad \Pi_1'\quad N(\mathrm{v})\wedge P(\mathrm{v})\quad \Pi_2\quad Q(\mathrm{v})}{}\ 3}{\nabla\mathrm{v}\,Q(\mathrm{v})}
$$

In Π', the track from hypothesis $\nabla\mathrm{v}\,M(\mathrm{v})$ to conclusion $\nabla\mathrm{v}\,Q(\mathrm{v})$ has the aspect:

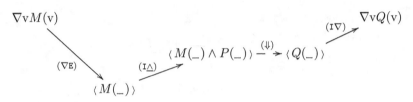

This track has 2 minimal formulas $(\langle M(\_)\rangle$ and $\langle Q(\_)\rangle)$ and 1 large formula $(\langle M(\_)\wedge P(\_)\rangle)$.

Theorem 2 (Growing systems). *Growing systems have the following properties.*

1. We can reduce growing rule applications to at most one central application of transformation: $\{(\mathrm{Elm}),(\mathrm{Tr}),(\mathrm{Int})\}^* \Rrightarrow (\mathrm{Elm})^*\,;\,(\mathrm{Int})^*\,;\,(\mathrm{Tr})^{\leq 1}\,;\,(\mathrm{Int})^*$. For marked rules, we have at most one final application of transformation:
   $\{(\mathrm{Elm}),(\mathrm{Tr}),(\mathrm{Int})\}^* \Rrightarrow (\mathrm{Int})^*\,;\,(\mathrm{Tr})^{\leq 1}$.

2. We can reduce growing marked tracks to one of the following forms.

   (0) *No application of transformation rule:* $\langle M\rangle\,(\mathrm{I}\bullet)^*\,\langle N\rangle$

$$
\langle M\rangle\ \nearrow\ \langle N\rangle
$$

   (<) *A single transformation before introductions:* $\langle M\rangle\,(\mathrm{Tr})\,\langle N\rangle\,(\mathrm{I}\bullet)^*\,\langle P\rangle$

$$
\langle M\rangle\ \dashrightarrow\ \langle N\rangle\ \nearrow\ \langle P\rangle
$$

($>$) *A single transformation after introductions:* $\langle M \rangle (\mathbf{I}\bullet)^*$
$\langle N \rangle (\underline{\text{Tr}}) \langle P \rangle$

$$\langle N \rangle \quad \dashrightarrow \quad \langle P \rangle$$
$$\nearrow$$
$$\langle M \rangle$$

3. *We can reduce growing derivation tracks to one of the following forms.*

   (0) *No application of transformation rule (1 minimal formula):*

$$H \qquad\qquad F$$
$$\searrow \qquad \nearrow$$
$$G \qquad\qquad\quad \text{minimal formula: } G$$

   ($<$) *A single* ($\underline{\text{Tr}}$) *before introductions (2 minimal marked formulas):*

$$H \qquad\qquad\qquad\qquad\qquad\qquad F$$
$$\searrow \qquad\qquad\qquad\qquad \nearrow$$
$$\langle N \rangle \quad \dashrightarrow \quad \langle P \rangle$$

$$\text{minimal formulas: } \langle N \rangle, \langle P \rangle$$

   ($>$) *A single* ($\underline{\text{Tr}}$) *after introductions (1 large and 2 minimal formulas):*

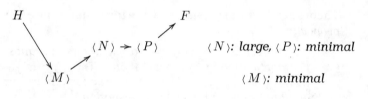

$$H \qquad\qquad\qquad\qquad F$$
$$\searrow \qquad\quad \langle N \rangle \rightarrow \langle P \rangle \qquad\qquad \langle N \rangle: \text{large}, \langle P \rangle: \text{minimal}$$
$$\searrow \nearrow$$
$$\langle M \rangle \qquad\qquad\qquad\qquad\qquad \langle M \rangle: \text{minimal}$$

Proof. By Corollary 2, Proposition 1 and Theorem 1.

$\square$

The first three logics, namely basic $\mathsf{BL}_\updownarrow$, proper $\mathsf{PL}_\updownarrow$ and upclosed $\mathsf{SL}_\downarrow$. are formulated without operational rules: we have *simple systems* $\mathsf{ND}(\Xi \cup \{(\underline{\text{Tr}})\})$, with $\Xi \subseteq \mathsf{NR} \cup \mathsf{PR}$ and $(\underline{\text{Tr}}) \in \{ (\updownarrow), (\Downarrow) \}$. We now examine such simple systems.

Theorem 3 (Simple systems). *A simple system has the following properties.*

1. *We can reduce rule applications to at most one central application of transformation:* $\{(\boxminus\mathsf{m}), (\underline{\text{Tr}}), (\mathsf{Int})\}^* \Rrightarrow$
   $(\boxminus\mathsf{m})^* ; (\mathsf{I\top})^{\leq 1} ; (\underline{\text{Tr}})^{\leq 1} ; (\bot\mathsf{E})^{\leq 1} ; (\mathsf{Int})^*$. *For marked rules:* $\{(\underline{\mathsf{Elm}}), (\underline{\text{Tr}}), (\underline{\mathsf{Int}})\}^*$
   $\Rrightarrow (\mathsf{I\top})^{\leq 1} ; (\underline{\text{Tr}})^{\leq 1} ; (\bot\mathsf{E})^{\leq 1}.$

2. *We can reduce simple marked tracks to one of the following forms.*

   (0) *No application of transformation rule:* $\top\,(\text{I}\underline{\top})\,\langle\top\rangle$ *or* $\langle\bot\rangle\,(\underline{\bot}\text{E})\,\bot$

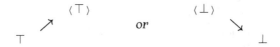

   (1) *A single application of transformation rule:* $\langle M\rangle\,(\underline{\text{Tr}})\,\langle N\rangle$

$$\langle M\rangle \quad \dashrightarrow \quad \langle N\rangle$$

3. *We can reduce simple derivation tracks to one of the following forms.*

   (0) *No application of transformation rule (1 minimal formula):*

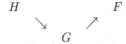

   minimal formula: G

   (1) *A single application of transformation (2 minimal marked formulas):*

$$
\begin{array}{ccc}
H & & F \\
\ \searrow & & \nearrow \\
& \langle N\rangle \quad \dashrightarrow \quad \langle P\rangle &
\end{array}
$$

   minimal formulas: $\langle N\rangle, \langle P\rangle$

Proof. By Corollary 2, Proposition 1 and Theorem 1.

$\square$

Corollary 3 (Sub-formula). *A simple system has the sub-formula property.*   ♭

Proof. By Theorem 3 (Simple systems).

$\square$

5   Concluding Remarks

We have considered 'generally' represented by generalized quantifiers and examined the structure of natural deduction derivations for some of its versions.

Table 12: Structure of reduced tracks

Logic rules	minimal	large	sub-formula
Basic BR	2	0	+
Proper PR	2	0	+
Upclosed SR	2	0	+
Lattice LR	2	1	$\pm$
Filter FR	2	1	$\pm$
Restricted RR	2	2	$\pm$
Other	m	n	$-$

The tracks in these logics can be reduced to the forms shown in Table 12.

A reduced track in restricted logics has at most one application of a transformation rule, whence it has at most 2 minimal and large formulas.

In basic, proper and upclosed logics, reduced tracks have no large formula, and these logics have the sub-formula property.

Restricted logics may fail to have the sub-formula property because of the presence of large formulas in tracks. But these large formulas limited to at most 2 are linked by a central application of transformation rule. For growing logics (like those for lattices and filters) one large formula is a sub-formula of the end formula F while the other one can be obtained from its minimal ones. For instance, the large formula $\langle P \rangle$ of a reduced track in FR is a marked conjunction of the minimal formulas of the introduction segments going to $\langle P \rangle$. In LR, such a large formula may be reduced to a marked disjunction of conjunctions, as we can reduce $(\langle M \vee N \rangle \wedge P)$ to $\langle (M \wedge P) \vee (N \wedge P) \rangle$.[19] So, even without the sub-formula property,

---

[19] We have a reduction $(I\underline{\vee})$; $(I\underline{\wedge}) \Rightarrow (I\underline{\wedge})$; $(I\underline{\vee})$; $(\updownarrow)$ as follows:

$$(I\underline{\wedge}) \dfrac{(I\underline{\vee}) \dfrac{\langle M \rangle \quad \langle N \rangle}{\langle M \vee N \rangle} \quad \langle P \rangle}{\langle M \vee N \rangle \wedge P \rangle}$$

$$\Downarrow$$

$$[(M \wedge P) \vee (N \wedge P)]^1$$

$$(\updownarrow) \dfrac{(I\underline{\vee}) \dfrac{(I\underline{\wedge}) \dfrac{\langle M \rangle \quad \langle P \rangle}{\langle M \wedge P \rangle} \quad (I\underline{\wedge}) \dfrac{\langle N \rangle \quad \langle P \rangle}{\langle N \wedge P \rangle}}{\langle (M \wedge P) \vee (N \wedge P) \rangle}}{\langle M \vee N) \wedge P \rangle} \qquad \dfrac{[M \vee N) \wedge P]^1}{} \ {}_1$$

Thus, we have: $\{ (I\underline{\vee}), (I\underline{\wedge}), (\updownarrow) \}^* \Rightarrow^* (I\underline{\wedge})^* ; (I\underline{\vee})^* ; (\updownarrow)^{\leq 1}$.

we still have some control on the structure of reduced derivations.

The results presented here refine those obtained before. In (Vana et al., 2014) we have considered modular formulations (with (⇕)). So, lattice and filter logics use (∧E). Here, we have introduced the new transformation rule (⇓) which behaves quite similarly to (⇕), and exploited it to gain finer control. Thus, formulations with (⇓), where appropriate, give sharper results.

References

Barwise, J. & Cooper, R. (1981), 'Generalized quantifiers and natural language', *Linguistics and Philosophy* 4, 159–219.

Carnielli, W. & Veloso, P. (1997), Ultrafilter logic and generic reasoning, *in* A. L. G. Gottlob & D. Mundici, eds, 'Computational Logic and Proof Theory', Vol. 1289 of *Lecture Notes in Computer Science*, pp. 34–53.

Dalen, D. v. (1989), *Logic and Structure*, Springer.

Mostowski, A. (1957), 'On a generalization of quantifiers', *Fundamenta Mathematicæ* 44, 12–36.

Prawitz, D. (1965), *Natural Deduction: a proof-theoretical study*, Wiksdell.

Vana, L. (2008), Dedução Natural e Cálculo de Seqüentes para 'Geralmente', PhD thesis, COPPE-UFRJ.

Vana, L., Veloso, P. & Veloso, S. (2005), Natural deduction strategies for 'generally', *in* 'Actas de la XI Conferencia de la Asociación Española para la Inteligencia Artificial', pp. 105–131.

Vana, L., Veloso, P. & Veloso, S. (2007), 'Natural deduction for 'generally'', *Logic Journal of the IGPL* 15(5), 775–800.

Vana, L., Veloso, P. & Veloso, S. (2014), *Advances in Natural Deduction*, Springer, chapter On the structure of natural Deduction derivations for 'generally', pp. 105–131.

Veloso, P. (2001), 'A logical approach to qualitative reasoning with 'several'', *Logique et Analyse* pp. 349–371.

Veloso, P. & Carnielli, W. (2004), *Logic, Epistemology and the Unity of Science*, Kluwer, chapter Logics for qualitative reasoning, pp. 487–526.

Veloso, P. & Veloso, S. (2001), On a logical framework for 'generally', *in* J. Abe & J. Silva Filho, eds, 'Logic, Artificial Intelligence and Robotics - Proc. LAPTEC 2001', pp. 279–286.

Veloso, P. & Veloso, S. (2004), *Aspects of Universal Logic*, Univ. Neuchâtel, chapter On modulated logics for 'generally': some meta-mathematical issues, pp. 146–168.

Veloso, P. & Veloso, S. (2005a), 'Functional interpretation of logics for 'generally'', *Logic Journal of the IGPL* **13**(1), 127–140.

Veloso, P. & Veloso, S. (2005b), 'On 'most' and 'representative': filter logic and special predicates', *Logic Journal of the IGPL* **13**(6), 717–728.

Veloso, P. & Veloso, S. (2011), *Logic without Frontiers*, College Publications, chapter Revisiting 'generally' and 'rarely', pp. 183–208.

Veloso, S. & Veloso, P. (2002), On special functions and theorem proving in logics for 'generally', *in* G. Bittencourt & G. Ramalho, eds, 'Advances in Artificial Intelligence: 16th Brazilian Symposium in Artificial Intelligence', Vol. 2507 of *Lecture Notes in Artificial Intelligence*, pp. 1–10.

Proof-theoretic validity based on elimination rules[‡]

Peter Schroeder-Heister[*]

[*] Wilhelm-Schickard-Institut für Informatik
Universität Tübingen
Sand 13, 72076 Tübingen, Germany
psh@uni-tuebingen.de

Abstract

In the tradition of Dummett-Prawitz-style proof-theoretic semantics, which considers validity of derivations or proofs as one of its core notions, this paper sketches an approach to proof-theoretic validity based on elimination rules and assesses its merits and limitations. Some remarks are made on alternative approaches based on the idea of dualizing connectives and proofs, as well as on definitional reflection using elimination clauses.

1 Introduction

Following Gentzen's dictum that "the introductions represent so-to-speak the 'definitions' of the corresponding signs" (Gentzen, 1934/35, p. 189), many approaches to proof-theoretic semantics (see Schroeder-Heister, 2012b) consider introduction rules to be basic, meaning giving, or self-justifying, whereas the elimination inferences are justified as valid with respect to the given introduction rules. The roots of this conception are threefold: First there is a verificationist theory according to which assertibility conditions of a sentence constitute its meaning. This seems to underly not only to a large extent the semantic conceptions of Dummett and

[‡]I am grateful to Luiz Carlos Pereira for many fruitful discussions on issues of proof theory and proof-theoretic semantics over more than 35 years. As the systematics of natural deduction and the delineation of intuitionistic logic has been a main theme of Luiz Carlos's work, the topic of this paper seems to me appropriate for a volume in his honour. I am grateful to Thomas Piecha, Dag Prawitz and the editors for many helpful comments and suggestions. The research reported here has been supported by the French-German ANR-DFG projects "Hypothetical Reasoning" and "Beyond Logic" (DFG Schr 275/16-2 and Schr 275/17-1).

Prawitz, which are the most developed ones in this respect, but the whole movement of intuitionism. Even if it is not directly connected to the verificationism of the early Wittgenstein and the Vienna circle, there are strong reminiscences of their positions in verificationist proof-theoretic semantics. There is a justificationist and verificationist bias in certain branches of constructive semantics and philosophy of mathematics. The second point is the idea that we must distinguish between what constitutes the meaning and what are the consequences of this meaning, in order to cope with the 'paradox of inference' (Cohen & Nagel, 1934, Ch. 9, § 1); see the discussion in Dummett (1975). For an inference to be informative, not every inference can be definitional. The informative inferences are established by reflection on the meaning of the expressions involved, without being meaning-constituting themselves. Whereas introduction steps are meaning giving, the remaining valid inferences give novel insight beyond what is 'definitionally' already contained in the premises. The third point, which is closely connected to the first, is the primacy of assertion over other speech acts such as assuming or denying, which is implicit in most approaches to proof-theoretic semantics. In Prawitz's definition of validity, and in intuitionistic semantics in general, assumptions are placeholders for proofs or constructions, and negation is reduced to implying absurdity. This yields a general bias towards positive forward reasoning, which is reflected in the primacy of forward-directed introductions (for a criticism of this approach see Schroeder-Heister, 2012a). To some extent this view is also implicit in the clause-based theory of definitional reflection (Hallnäs, 1991; Schroeder-Heister, 1993), as clauses are directed from bodies to heads, that is, from defining conditions towards defined atoms. The non-determinism in clauses, i.e., the fact that several clauses may define the same atom (which in logic we have, for example, with the introduction rules for disjunction) emphasizes this directedness. Whereas the use of a single clause is simply the application of a definition, definitional reflection extracts additional content from an expression with respect to the definition as a whole, which can be viewed as generating valid informative inferences.

The division between introductions and eliminations suggests to exchange their roles and thus to consider elimination rules rather than introduction rules as the basis of proof-theoretic semantics. Such an approach would be nearer to a falsificationist methodology in Popper's sense. The philosophical problems and shortcomings of verificationism, which cannot be discussed here, would be strong arguments in favour of this alternative. The second point mentioned in the previous paragraph is indifferent with respect to the primacy of introduction or elimination rules, as it only says that there must be *one part* of the rules which is meaning giving and *another one* which is informative, so one may as well choose the elimination rules as meaning giving. The third point, the primacy of assertion, would be replaced with the Popperian claim that conjectures and therefore assumptions are primary to assertions.

This possibility has been intensively discussed by the main advocates of verificationist semantics. In Dummett it often runs under the heading of a 'pragmatist' theory of meaning and has received considerable credit in some of his publications such as Dummett (1976)[1]. Some technical ideas towards a proof-theoretic semantics based on elimination rather than introduction rules have been sketched by Dummett (1991, Ch. 13). A precise definition of validity based on elimination inferences is due to Prawitz (1971, 2007). In slightly improved form, it will be presented in Section 3. As its background, we recall the validity definition based on introduction rules.

2 The introductions-based approach: Derivation structures, justifications and atomic systems

The definition of validity refers to a general notion of derivation structures and reductions that justify derivations, as well as to atomic systems. We refer to the version that is given in Schroeder-Heister (2006, 2012b), which is an interpretation of Prawitz's notion of proof-theoretic validity. We consider only the constants of positive propositional logic (conjunction, disjunction, implication). We assume that an atomic system S is given as determining the derivability of atomic formulas, which is the same as their validity. A formula over S is a formula built up by means of logical connectives starting with atoms from S. We want to define the validity of a derivation which proceeds from formulas over S as assumptions to a formula over S as conclusion. Such a derivation is not necessarily a derivation in a given formal system: We want to tell of an *arbitrary* derivation whether it is valid or not. We propose the term "derivation structure" for such an arbitrary derivation. (Prawitz uses various terminologies, such as "[argument or proof] schema" or "[argument or proof] skeleton".) Derivation structures are candidates for valid derivations. More precisely, a derivation structure is a formula tree which resembles a natural deduction tree with the difference that it is composed of arbitrary rules. Such rules can have arbitrary and arbitrarily many premisses, and each

---

[1]However, the main theses of this paper were withdrawn in the preface to Dummett (1993).

premiss may depend on assumptions which are discharged at this step. So the general form of an inference rule is the following, where the square brackets indicate assumptions which can be discharged at the application of the rule:

$$\frac{\begin{array}{ccc}[C_{11},\ldots,C_{1m_1}] & & [C_{n1},\ldots,C_{nm_n}] \\ A_1 & \cdots & A_n\end{array}}{B}, \quad \text{in short:} \quad \frac{\begin{array}{ccc}[\Gamma_1] & & [\Gamma_n]\\ A_1 & \cdots & A_n\end{array}}{B}.$$

Obviously, the standard introduction and elimination rules are particular cases of such rules. As a generalization of the standard reductions of maximal formulas it is supposed that certain reduction procedures are given. A reduction procedure transforms a given derivation structure into another one. A set of reduction procedures is called a *derivation reduction system* and denoted by $\mathcal{J}$. Reductions serve as justifying procedures for non-canonical steps. These are steps, which are not self-justifying, i.e., which are not introduction steps. Therefore a reduction system $\mathcal{J}$ is also called a *justification*. Reduction procedures must satisfy certain constraints such as closure under substitution. As the validity of a derivation not only depends on the atomic system S but also on the derivation reduction system used, we define the validity of a derivation structure with respect to the underlying atomic basis S and with respect to the justification $\mathcal{J}$. A *canonical* derivation structure is a derivation structure which uses an introduction rule in the last step. It is called *open*, if it depends on undischarged assumptions, otherwise it is called *closed*.

Definition: Validity based on introduction rules

(i) Every closed derivation in S is S-valid with respect to $\mathcal{J}$ (for every $\mathcal{J}$).

(ii) A closed canonical derivation structure is S-valid with respect to $\mathcal{J}$, if its immediate substructure $\begin{array}{c}A\\ \mathcal{D}\\ B\end{array}$ is S-valid with respect to $\mathcal{J}$.

(iii) A closed non-canonical derivation structure is S-valid with respect to $\mathcal{J}$, if it reduces, with respect to $\mathcal{J}$, to a canonical derivation structure, which is S-valid with respect to $\mathcal{J}$.

(iv) An open derivation structure $\begin{array}{c}A_1\ldots A_n\\ \mathcal{D}\\ B\end{array}$, where all open assumptions of $\mathcal{D}$ are among $A_1,\ldots,A_n$, is S-valid with respect to $\mathcal{J}$, if for *every extension S' of S and every extension $\mathcal{J}'$ of $\mathcal{J}$, and for every list of closed derivation structures $\dfrac{\mathcal{D}_i}{A_i}$ $(1 \leq i \leq n)$, which are*

S'-valid with respect to $\mathcal{J}'$, $\quad \dfrac{\begin{array}{ccc} \mathcal{D}_1 & & \mathcal{D}_n \\ A_1 & \cdots & A_n \\ & \mathcal{D} & \end{array}}{B}$ *is S'-valid with respect to $\mathcal{J}'$.*

(See Prawitz, 1973, p. 236; 1974.; p. 73; 2006).

In clause (iv), the reason for considering extensions $\mathcal{J}'$ of $\mathcal{J}$ and of extensions S' of S is a monotonicity constraint. Derivations should remain valid if one's knowledge incorporated in the atomic system and in the reduction procedures is extended. The consideration of such extensions, which can be found in Prawitz (1971), is a point of deviation of this exposition from later definitions given by Prawitz.

The S-validity of a generalized inference rule

$$\dfrac{\begin{array}{ccc} [\Gamma_1] & & [\Gamma_n] \\ A_1 & \cdots & A_n \end{array}}{B}$$

with respect to a justification $\mathcal{J}$ means that for all derivations $\dfrac{\begin{array}{ccc} \Gamma_1 & & \Gamma_n \\ \mathcal{D}_1 & & \mathcal{D}_n \\ A_1 & & A_n \end{array}}{}, \dots,$ which are S'-valid with respect to $\mathcal{J}'$ for extensions S' and $\mathcal{J}'$ of S and $\mathcal{J}$, respectively, the derivation

$$\dfrac{\begin{array}{ccc} (1) & & (1) \\ [\Gamma_1] & & [\Gamma_n] \\ \mathcal{D}_1 & & \mathcal{D}_n \\ A_1 & \cdots & A_n \end{array}}{B} \, (1)$$

is S'-valid with respect to $\mathcal{J}'$. For a simple inference rule

$$\dfrac{A_1 \, \cdots \, A_n}{A}$$

this means that it is valid with respect to $\mathcal{J}$, if the one-step derivation structure of the same form is S-valid with respect to $\mathcal{J}$.

This gives rise to a corresponding notion of consequence (see also Prawitz, 1985). Instead of saying that the rule

$$\dfrac{A_1 \quad \cdots \quad A_n}{A}$$

is S-valid with respect to $\mathcal{J}$, we may say that A is a *consequence* of $A_1, \ldots, A_n$ *with respect to* S *and* $\mathcal{J}$ ($A_1, \ldots, A_n \models_{S,\mathcal{J}} A$). If this holds for any S, we may speak of *universal consequence with respect to* $\mathcal{J}$ ($A_1, \ldots, A_n \models_{\mathcal{J}} A$); and finally, if there is some $\mathcal{J}$ such that we have universal consequence with respect to $\mathcal{J}$, then we may speak of *logical consequence* ($A_1, \ldots, A_n \models A$).

If for $\mathcal{J}$ we choose the standard reductions of intuitionistic logic, then all derivations in intuitionistic logic are valid with respect to $\mathcal{J}$, thus establishing the *soundness* of intuitionistic logic with respect to introductions-based proof-theoretic semantics. We may ask if the converse holds, namely whether, given that a derivation $\mathcal{D}$ is valid with respect to some $\mathcal{J}$, there is a derivation in intuitionistic logic with the same end-formula and without any open assumptions beyond those already open in $\mathcal{D}$. That intuitionistic logic is *complete* in this sense has been conjectured by Prawitz (see Prawitz, 1973, 2014). This conjecture is not without problems as results by Sandqvist (2009) and Piecha et al. (2014) indicate.

3   Validity based on elimination rules

In the approach based on elimination rules, the elimination inferences are considered 'self-justifying', and the introduction rules are justified with respect to them. The reductions need not to be changed for that purpose. The standard reductions for the logical constants can serve for the justification of the introductions from the eliminations as well. However, additional reductions must be considered which correspond to the permutative reductions in natural deduction. In this section, we speak of validity$_E$ as validity based on elimination rules in contradistinction to validity$_I$ as validity based on introduction rules.

The idea behind validity$_E$ is that, if all applications of elimination rules to the complex end-formula A of a derivation structure $\mathcal{D}$ yield S-valid$_E$ derivation structures or reduce to such (with respect to a justification $\mathcal{J}$), then $\mathcal{D}$ is itself S-valid$_E$ (with respect to $\mathcal{J}$). This suggests the following definition for positive propositional logic:

Definition: Validity based on elimination rules

(i) *Every closed derivation in S is S-valid$_E$ with respect to $\mathcal{J}$ (for every $\mathcal{J}$).*

(ii∧) *A closed derivation structure* $\begin{array}{c} \mathcal{D} \\ \hline A \wedge B \end{array}$ *is S-valid$_E$ with respect to $\mathcal{J}$,* if the closed derivation structures $\dfrac{\mathcal{D}}{\dfrac{A \wedge B}{A}}$ and $\dfrac{\mathcal{D}}{\dfrac{A \wedge B}{B}}$ are S-valid$_E$ with respect to $\mathcal{J}$, or reduce to derivation structures, which are S-valid$_E$ with respect to $\mathcal{J}$.

(ii →) A closed derivation structure $\dfrac{\mathcal{D}}{A \to B}$ is S-valid$_E$ with respect to $\mathcal{J}$, if for every extension S' of S and for every extension $\mathcal{J}'$ of $\mathcal{J}$, and for every closed derivation structure $\dfrac{\mathcal{D}'}{A}$, which is S'-valid$_E$ with respect to $\mathcal{J}'$, the (closed) derivation structure $\dfrac{\dfrac{\mathcal{D}}{A \to B} \quad \dfrac{\mathcal{D}'}{A}}{B}$ is

S'-valid$_E$ with respect to $\mathcal{J}'$, or reduces to a derivation structure, which is S'-valid$_E$ with respect to $\mathcal{J}'$.

(ii∨) A closed derivation structure $\dfrac{\mathcal{D}}{\dfrac{A \vee B}{A B}}$ is S-valid$_E$ with respect to $\mathcal{J}$, if for every extension S' of S and every extension $\mathcal{J}'$ of $\mathcal{J}$, and for all derivation structures $\mathcal{D}_1$ and $\mathcal{D}_2$ with atomic C, which are $\dfrac{C}{}$ $\dfrac{C}{}$ S'-valid$_E$ with respect to $\mathcal{J}'$ and which depend on no assumptions beyond A and B, respectively, the (closed) derivation struc-

ture $\dfrac{\dfrac{\mathcal{D}}{A \vee B} \quad \dfrac{[A]^{(1)}}{\dfrac{\mathcal{D}_1}{C}} \quad \dfrac{[B]^{(1)}}{\dfrac{\mathcal{D}_2}{C}}}{C}{}^{(1)}$ is S'-valid$_E$ with respect to $\mathcal{J}'$, or

reduces to a derivation structure, which is S'-valid$_E$ with respect to $\mathcal{J}'$.

(iii) A closed derivation structure $\dfrac{\mathcal{D}}{A}$ of an atomic formula A, which is not a derivation in S, is S-valid$_E$ with respect to $\mathcal{J}$, if it reduces with respect to $\mathcal{J}$ to a derivation in S.

(iv) An open derivation structure $\dfrac{A_1 \ldots A_n}{\dfrac{\mathcal{D}}{B}}$, where all open assumptions of $\mathcal{D}$ are among $A_1, \ldots, A_n$, is S-valid$_E$ with respect to $\mathcal{J}$, if for every extension S' of S and every extension $\mathcal{J}'$ of $\mathcal{J}$, and for every list of closed derivation structures $\dfrac{\mathcal{D}_i}{A_i}$ $(1 \leq i \leq n)$, which are

S'-valid$_E$ with respect to $\mathcal{J}'$,
$$\begin{array}{cc} \mathcal{D}_1 & \mathcal{D}_n \\ A_1 & \cdots \quad A_n \\ & \mathcal{D} \\ & B \end{array}$$
is S'-valid$_E$ with respect to $\mathcal{J}'$.

Clause (iv) is identical with clause (iv) in the definition of introductions-based validity in Section 2, i.e., open assumptions in derivations are interpreted in the same way as before, namely as placeholders for closed valid derivations. Note that clause (iii) is needed, as we do not have here the notion of a canonical derivation. In the definition of validity based on introduction rules, the case considered in clause (iii) was a special case of non-canonical derivations. Clauses (i) and (iii) can be conjoined to form the single clause

(i/iii) *A closed derivation structure* $\begin{array}{c} \mathcal{D} \\ A \end{array}$ *of an atomic formula A is S-valid$_E$ with respect to $\mathcal{J}$, if it reduces with respect to $\mathcal{J}$ to a derivation in S.*

The validity$_E$ of an inference rule as well as the notions of consequence and logical consequence are defined exactly as in the introductions-based approach of Section 2.

It is crucial that the minor premisses C in the application of $\vee$-elimination (and similarly for $\exists$-elimination, if we deal with quantifiers) are atomic, otherwise the induction over the end-formulas of derivations, on which this definition is based, would break down. Prawitz (1971, Appendix A.2), eliminations-based definition of validity was without clauses for disjunction (and existential quantification), as he had not been aware at the time that for the purpose of defining validity the restriction to atomic C is sufficient (repeated in Schroeder-Heister, 2006). The revised proposal with atomic C was published in Prawitz (2007). There he refers to the fact that also Dummett (1991, Ch. 13), in his remarks on a "pragmatist" theory of meaning with an inverse justification based on elimination rules uses atomic C. The fact that one can do without complex C is closely related to the fact that the definability of first-order logical constants in second-order propositional $\forall \rightarrow$-logic, which was first observed by Prawitz (1965, Ch. 5), can already be obtained in *predicative* second-order $\forall \rightarrow$-logic in the sense that the latter proves the introduction and elimination rules for the defined connectives as shown by Ferreira (2006, see also Ferreira & Ferreira 2013)[2].

The condition "or reduces to a derivation structure, which is S'-valid$_E$ with respect to $\mathcal{J}'$" at the end of clauses (ii→), (ii∧), (ii∨) is called the 'reduction condition'. It corresponds to the basic intuition of proof-theoretic validity semantics that a derivation is valid, if it is of a certain form *or reduces to such a form*. It simplifies certain proofs such as that of the validity$_E$ of the introduction inferences. However, it can be omitted without loss of definitional power, since both with and without the reduction condition we can show that a derivation structure is valid$_E$ if and only if it reduces to a valid$_E$ derivation structure. In fact, in the original notion of validity$_E$ envisaged by Dummett and defined by Prawitz (and also in corresponding notions of computability) the notion of reduction does not come in until the atomic stage is reached. In any case a reduction condition must be contained in clause (iii) which governs derivations of atomic sentences.

The standard reductions, which remove maximum formulas, are not sufficient to show that all introduction and elimination rules are valid. Due to the restriction that C must be atomic, we now have to justify in particular the ∨-elimination rule for *nonatomic* C. For that we need reductions, which correspond to permutative reductions in natural deduction. For example, in order to show that

$$
\cfrac{\mathcal{D} \quad \cfrac{\overset{(1)}{[A]}}{\underset{\mathcal{D}_1}{C_1 \wedge C_2}} \quad \cfrac{\overset{(1)}{[B]}}{\underset{\mathcal{D}_2}{C_1 \wedge C_2}}}{C_1 \wedge C_2} \; (1)
$$

is valid, given that $\mathcal{D}$, $\mathcal{D}_1$ and $\mathcal{D}_2$ are valid, we need to use reductions according to which

$$
\cfrac{\cfrac{\mathcal{D} \quad \cfrac{\overset{(1)}{[A]}}{\underset{\mathcal{D}_1}{C_1 \wedge C_2}} \quad \cfrac{\overset{(1)}{[B]}}{\underset{\mathcal{D}_2}{C_1 \wedge C_2}}}{C_1 \wedge C_2}\;(1)}{C_i}
\quad \text{reduces to} \quad
\cfrac{\mathcal{D} \quad \cfrac{\cfrac{\overset{(1)}{[A]}}{\underset{\mathcal{D}_1}{C_1 \wedge C_2}}}{C_i} \quad \cfrac{\cfrac{\overset{(1)}{[B]}}{\underset{\mathcal{D}_2}{C_1 \wedge C_2}}}{C_i}}{C_i}\;(1)\;.
$$

---

[2]This fact was independently discovered by Sandqvist.

The rule of importation

$$(R_{imp}) \quad \frac{A \to (B \to C)}{A \land B \to C}$$

is an instructive example to compare the justifications based on validity$_I$ vs. validity$_E$. For validity$_E$, we would now need the following reduction as a justification:

$$\cfrac{\cfrac{\cfrac{\mathcal{D}}{A \to (B \to C)}}{A \land B \to C} \quad \cfrac{\mathcal{D}'}{A \land B}}{C} \quad \text{reduces to} \quad \cfrac{\cfrac{\mathcal{D}}{A \to (B \to C)} \quad \cfrac{\cfrac{\mathcal{D}'}{A \land B}}{A}}{\cfrac{B \to C}{C}} \quad \cfrac{\mathcal{D}'}{A \land B}$$

We do not need any of the standard reductions, which means that importation is valid with respect to the justification consisting of this reduction alone.

In order to justify (R_{imp}) with respect to validity$_I$, we would rely on a similar reduction: $\cfrac{\mathcal{D}}{\cfrac{A \to (B \to C)}{A \land B \to C}}$

$$\text{reduces to} \quad \cfrac{\cfrac{\cfrac{\mathcal{D}}{A \to (B \to C)} \quad \cfrac{[A \land B]^{(1)}}{A}}{\cfrac{B \to C}{C}} \quad \cfrac{[A \land B]^{(1)}}{B}}{A \land B \to C}{}^{(1)}$$

However, we would need to use in addition the standard reductions of conjunction and implication in order to justify the $(\land \text{E})$ and $(\to \text{E})$ steps involved (see the supplement to Schroeder-Heister, 2012b). In both cases we must use a reduction that unfolds a single step into a succession of more elementary steps.

It is not entirely clear which logic we obtain by the eliminations-based approach. From the remarks above it is clear that we can justify the rules of intuitionistic logic, which means that intuitionistic logic is sound with respect to this semantics. In view of our definitional clause for disjunction, it is natural to consider atomic second-order propositional logic $\mathbf{F_{at}}$ as the formal system corresponding to eliminations-based semantics. This system, in which $A \lor B$ is interpreted as $\forall X((A \to X) \to ((B \to X) \to X))$, where the universal quantifier runs over atomic propositions only, has been studied by Ferreira (2006). Though it does not contain disjunction as a primitive sign, it satisfies the disjunction property for the second-order interpretation of disjunction. In fact, Ferreira & Ferreira (2014) could show that it is equivalent to intuitionistic logic. However, whether $\mathbf{F_{at}}$ and thus intuitionistic logic is complete with respect to eliminations-based semantics is not obvious and can be questioned. As mentioned before, there are arguments that problematize Prawitz's completeness conjecture for intuitionistic logic with respect to introductions-based proof-theoretic semantics. Depending on certain assumptions about the form of atomic systems and on the way of dealing with hypothetical proofs[3], there are actually counterexamples to completeness (see Piecha et al., 2014; Piecha, 2015). These counterexamples can be adapted to the eliminations-based approach, as the handling of atomic systems and hypothetical proofs does not differ between the two approaches.

It should be remarked that there is a notion of computability based on elimination rules used in proofs of (strong) normalization which corresponds to the eliminations-based notion of validity. Actually, this notion is more common in presentations of this topic than computability based on introductions-based(see, for example, Troelstra & Schwichtenberg, 2000, Ch. 6.6).

4   Co-implication and other alternatives

The intuition behind the approach based on elimination rules is that a derivation is valid, if the result of the application of each possible elimination rule to its end-formula is valid. This means that even a closed derivation is not valid due to its actual form or to the form to which it can be reduced (as in the introductions-based approach), but due to appending further inference steps to it. Its validity depends on that of the immediate consequences we can reach starting with this derivation. So one might call it a consequentialist view of validity. This is an original approach, which brings a fresh idea into proof-theoretic validity. It must be noted, however, that basic tenets of introductions-based validity concepts are kept. Among those is the primacy of closed derivatons and the interpretation of open derivations. In both validity conceptions the definition of validity starts with closed derivations. And in both conceptions the validity of an open derivation is defined via the substitution of closed derivations for the open assumptions in open derivations, as expressed by the fact that clause (iv) of the definition of validity, which deals with open derivations, is identical

---

[3]All counterexamples assume that we consider arbitrary *extensions* of atomic systems when interpreting hypothetical proofs — an assumption now longer made by Prawitz.

in both of them. This means that both are biased towards assertions (by means of closed derivations), whereas assumptions are just placeholders for what can be asserted by means of closed derivations. It is assertions which, in the eliminations-based approach, are justified by their consequences. It is definitely not the case that assumptions receive a stronger stance in this sort of theory.

Therefore the approach sketched here is not the only possible and perhaps not even the most genuine way of putting elimination rules first. An eliminations-based approach which reverses the conceptual priority between assertions and assumptions would be one which considers derivations from assumptions to be primary. Such an approach can be obtained by dualizing the introductions-based approach by putting "deriving from" rather than "deriving of" in front. One would then develop ideas such as the following: A *closed derivation from A* should be a derivation of absurdity from A (corresponding to the fact that a closed derivation in the standard conception can be viewed as a derivation from truth), and a derivation $\begin{array}{c} A \\ \mathcal{D} \\ B \end{array}$ should be justified, if, for every closed valid derivation $\begin{array}{c} B \\ \mathcal{D}' \end{array}$ from B, $\begin{array}{c} A \\ \mathcal{D} \\ B \\ \mathcal{D}' \end{array}$ is a closed valid derivation from A. A full dualization would even lead to some variant of a single-assumption/multiple-conclusion logic, whose derivations are branching downwards rather than upwards. A closed derivation from A, in which all downward branches end with absurdity, might be called a closed refutation of A. If one of these branches ends with a formula B different from absurdity, it is an open refutation of A in the sense that replacing B with a closed refutation of B yields a closed refutation of A. Such approaches would lead to rules for logical constants which are dual to the standard ones. Conjunction (as the dual of disjunction) would be the constant that is canonically refuted by a refutation of A as well as by one of B, disjunction (as the dual of conjunction) would be the constant that is canonically refuted by a refutation of both A and B etc. Co-implication would come in as the dual of implication, which is canonically refuted by an open refutation of B to A, that is, of B given a refutation of A, and so on. This leads essentially to an approach in which usual derivation trees are written upside down, the concept of derivation is interchanged with that of refutation, etc. It corresponds to a system of dual-intuitionistic logic, in which connectives are replaced with their duals, and in particular implication by co-implication. However, structurally, the standard approach and its dual are the same — writing derivations upside down is not really an essential change. So if we want to obtain any conceptual gain from the consideration of dual concepts, we should be able to develop a joint system for both notions. A genuine elimination-rule approach might be desirable if one wanted to logically elaborate ideas like Popper's falsificationism by establishing refutation as the basis of reasoning. However, it is still not entirely clear what such an approach, which was already discussed by Popper (1948) and whose proof theory was initiated by Prawitz (1965, Appendix B.2), should look like formally. It would be a justification of what is now called 'bi-intuitionistic logic', which incorporates both implication and co-implication (see Wansing, 2008, 2013; Tranchini, 2012; Kapsner, 2014, and the references therein).[4]

Under normal circumstances, multiple-conclusion proof systems go beyond intuitionistic logic. However, by means of certain restrictions concerning the dependencies between formulas, constructivity in the intuitionistic sense can be enforced. Such systems have been studied by de Paiva & Pereira (1995, 2005). It would be a worthwhile task to turn this idea into some form of proof-theoretic semantics, which keeps track of such dependencies in the form of semantic conditions.

5 Consequentialism and definitional reflection

Definitional reflection adapts basic ideas concerning harmony and inversion from the logical realm to the realm of clausal definitions of atoms, inspired by a proof-theoretic interpretation of logic programming (Hallnäs, 1991; Schroeder-Heister, 1993, 2012b). In the simplest case, a definition is a finite list of clauses of the form

$$b_1, \ldots, b_n \Rightarrow a$$

where $b_1, \ldots, b_n, a$ are atoms. A finite set of clauses with the same head a

$$\left\{ \begin{array}{c} b_{11}, \ldots, b_{1n_1} \Rightarrow a \\ \vdots \\ b_{k1}, \ldots, b_{kn_k} \Rightarrow a \end{array} \right.$$

is called a *definition* of a. Then the rules of *definitional closure* says that we may pass along any definitional clause from its body to its head,

---

[4]A proof-theoretic approach, which does not consider co-implication but mixes standard implication with conjunction and disjunction (which are dual to one another) is hinted at in Dummett (1991, Ch. 13), and has been worked out (and improved) in detail by Litland (2012).

yielding rules

$$\frac{b_{i1} \ \ldots \ b_{in_1}}{a} \quad \ldots \quad \frac{b_{k1} \ \ldots \ b_{kn_k}}{a}$$

which correspond to introduction rules in logic. The powerful rule of *definitional reflection* says that anything that can be obtained from each defining condition of a can be obtained from a itself, which corresponds to the idea of elimination inferences:

$$\frac{a \qquad \overset{[b_{11},\ldots,b_{1n_1}]}{C} \qquad \ldots \qquad \overset{[b_{k1},\ldots,b_{kn_k}]}{C}}{C} \ .$$

Even though the rules of definitional closure and reflection come as a pair, without any of them primary over the other, there is some implicit bias towards introductions since clauses are directed. Definitional closure is interpreted as expressing the direction from definiens to definiendum, and definitional reflection as expressing the opposite direction. Changing this bias and inverting it, would have to be a radical reform of what a definition looks like. We would then have to consider 'consequential' clauses which determine the consequences of a given atom, such as

$$\left\{ \begin{array}{l} a \Rightarrow b_1 \\ \quad \vdots \\ a \Rightarrow b_m \end{array} \right. \ .$$

Definitional closure would then express reasoning along these consequential clauses

$$\frac{a}{b_1} \quad \ldots \quad \frac{a}{b_m}$$

which correspond to elimination inferences, and definitional reflection would be an introduction rule telling that a can be introduced from all possible definitional consequences taken together

$$\frac{b_1 \ \ldots \ b_m}{a} \ .$$

To make this approach reasonably expressive, we would have to consider also complex conclusions of consequential clauses rather than just atoms b_i. A multiple-conclusion clause

$$a \ \Rightarrow c_1,\ldots,c_n$$

would then be interpreted by a reflection rule like

$$\frac{\begin{array}{cccc} & [c_1] & & [c_n] \\ a & C & \cdots & C \end{array}}{C}.$$

If we have more than one multiple-conclusion clause, we would have to consider an appropriate list of reflection (elimination) rules. Alternatively, we could just consider single clauses, but with structural implications (rules) as conclusion, be means of which we can code in principle what can be expressed, e.g., by disjunction. Such a clause would, e.g. be

$$a \Rightarrow ((c_1 \Rightarrow p), \ldots, (c_n \Rightarrow p) \Rightarrow p)$$

for a fresh variable p. This leads to an eliminations-based approach for atomic rules corresponding to the one discussed in Section 3 for the case of logical constants. It dualizes the notion of a definition, but not the concept of a derivation, which is, as before, a concept of derivation of single formulas from (possibly) multiple assumptions. A complete dualization using single-assumption / multiple-conclusion derivations would trivialize the whole notion by just exchanging the right and left sides of definitions and the assumptions and conclusions of derivations, as does the dual-intuitionistic approach mentioned in Section 4. Using multiple-assumptions / multiple-conclusion derivations based on both standard definitional clauses and consequential ones should lead to an atomic framework corresponding to bi-intuitionistic logic. Such a theory still needs to be worked out. Particular attention should be paid to the question, which connections to logic programming and inductive definitions remain, and, therefore, how far definitional reasoning keeps the computational content which is inherent in forward-directed clauses.[5]

References

Cohen, M. R. & Nagel, E. (1934), *An Introduction to Logic and Scientific Method*, Routledge and Kegan Paul, London. First part reprinted as *An Introduction to Logic*, Harcourt, Brace and Co., 1962, 2nd ed. Hackett, Indianapolis, 1993.

de Paiva, V. & Pereira, L. C. (1995), 'A new proof system for intuitionistic logic (Abstract)', *Bulletin of Symbolic Logic* **1**, 101.

---

[5]Some remarks on dual frameworks can be found in Schroeder-Heister (2011).

de Paiva, V. & Pereira, L. C. (2005), 'A short note on intuitionistic propositional logic with multiple conclusions', *Manuscrito: Revista Internacional de Filosofia* **28**, 317–329.

Dummett, M. (1975), 'The justification of deduction', *Proceedings of the British Academy* pp. 201–232. Separately published by the British Academy 1973. Reprinted in Dummett, M.: Truth and Other Enigmas, London: Duckworth 1978.

Dummett, M. (1976), What is a theory of meaning? II, *in* G. Evans & J. McDowell, eds, 'Truth and Meaning', Oxford University Press, Oxford, pp. 67–137. Reprinted in Dummett, M.: The Seas of Language, Oxford: Clarendon Press 1993.

Dummett, M. (1991), *The Logical Basis of Metaphysics*, Duckworth, London.

Dummett, M. (1993), *The Seas of Language*, Clarendon Press, Oxford.

Ferreira, F. (2006), 'Comments on predicative logic', *Journal of Philosophical Logic* **35**, 1–8.

Ferreira, F. & Ferreira, G. (2013), 'Atomic polymorphism', *Journal of Symbolic Logic* **78**, 260–274.

Ferreira, F. & Ferreira, G. (2014), The faithfulness of atomic polymorphism, *in* A. Indrzejczak, J. Kaczmarek & M. Zawidzki, eds, 'Trends in Logic XIII: Gentzen's and Jaśkowski's Heritage — 80 Years of Natural Deduction and Sequent Calculi', Łódź University Press, pp. 55–65.

Gentzen, G. (1934/35), 'Untersuchungen über das logische Schließen', *Mathematische Zeitschrift* **39**, 176–210, 405–431 (English translation in: *The Collected Papers of Gerhard Gentzen* (ed. M. E. Szabo), Amsterdam: North Holland (1969), pp. 68–131).

Hallnäs, L. (1991), 'Partial inductive definitions', *Theoretical Computer Science* **87**, 115–142.

Kapsner, A. (2014), *Logics and Falsifications*, Springer, Berlin.

Litland, J. (2012), *Topics in Philosophical Logic. Ph.D. Thesis.*, Harvard University, Cambridge Mass.

Piecha, T. (2015), Completeness in proof-theoretic semantics, *in* T. Piecha & P. Schroeder-Heister, eds, 'Advances in Proof-Theoretic Semantics', Springer, Dordrecht.

Piecha, T., de Campos Sanz, W. & Schroeder-Heister, P. (2014), 'Failure of completeness in proof-theoretic semantics', *Journal of Philosophical Logic*. Online Aug 2014.

Popper, K. R. (1948), 'On the theory of deduction. I: Derivation and its generalizations. II: The definitions of classical and intuitionist negation', *Indagationes Mathematicae* **10**, 44—54, 111–120.

Prawitz, D. (1965), *Natural Deduction: A Proof-Theoretical Study*, Almqvist & Wiksell, Stockholm. Reprinted Mineola NY: Dover Publ., 2006.

Prawitz, D. (1971), Ideas and results in proof theory, *in* J. E. Fenstad, ed., 'Proceedings of the Second Scandinavian Logic Symposium (Oslo 1970)', North-Holland, Amsterdam, pp. 235–308.

Prawitz, D. (1973), Towards a foundation of a general proof theory, *in* P. Suppes et al., ed., 'Logic, Methodology and Philosophy of Science IV', North-Holland, pp. 225–250.

Prawitz, D. (1974), 'On the idea of a general proof theory', *Synthese* **27**, 63–77.

Prawitz, D. (1985), 'Remarks on some approaches to the concept of logical consequence', *Synthese* **62**, 152–171.

Prawitz, D. (2006), 'Meaning approached via proofs', *Synthese* **148**, 507–524. Special issue *Proof-Theoretic Semantics*, edited by R. Kahle and P. Schroeder-Heister.

Prawitz, D. (2007), Pragmatist and verificationist theories of meaning, *in* R. E. Auxier & L. E. Hahn, eds, 'The Philosophy of Michael Dummett', Open Court, Chicago, pp. 455–481.

Prawitz, D. (2014), An approach to general proof theory and a conjecture of a kind of completeness of intuitionistic logic revisited, *in* L. C. Pereira, E. H. Haeusler & V. de Paiva, eds, 'Advances in Natural Deduction: A Celebration of Dag Prawitz's Work', Springer, Dordrecht, pp. 269–279.

Sandqvist, T. (2009), 'Classical logic without bivalence', *Analysis* **69**, 211–218.

Schroeder-Heister, P. (1993), Rules of definitional reflection, *in* 'Proceedings of the 8th Annual IEEE Symposium on Logic in Computer Science (Montreal 1993)', IEEE Press, Los Alamitos, pp. 222–232.

Schroeder-Heister, P. (2006), 'Validity concepts in proof-theoretic semantics', *Synthese* **148**, 525–571. Special issue *Proof-Theoretic Semantics*, edited by R. Kahle and P. Schroeder-Heister.

Schroeder-Heister, P. (2011), Schluß und Umkehrschluß: Ein Beitrag zur Definitionstheorie, *in* C. F. Gethmann, ed., 'Lebenswelt und Wissenschaft. Kolloquienbeiträge und öffentliche Vorträge des XXI. Deutschen Kongresses für Philosophie (Essen 2008)', Meiner, Hamburg.

Schroeder-Heister, P. (2012a), 'The categorical and the hypothetical: A critique of some fundamental assumptions of standard semantics', *Synthese* **187**, 925–942. Special issue *The Philosophy of Logical Consequence and Inference*, edited by S. Lindström, E. Palmgren and D. Westerståhl.

Schroeder-Heister, P. (2012b), Proof-theoretic semantics, *in* E. Zalta, ed., 'Stanford Encyclopedia of Philosophy', http://plato.stanford.edu, Stanford.

Tranchini, L. (2012), 'Natural deduction for dual intuitionistic logic', *Studia Logica* **100**, 631–648.

Troelstra, A. S. & Schwichtenberg, H. (2000), *Basic Proof Theory*, Cambridge University Press (second edition).

Wansing, H. (2008), 'Constructive negation, implication, and co-implication', *Journal of Applied Non-Classical Logics* **18**, 341–364.

Wansing, H. (2013), 'Falsification, natural deduction and bi-intuitionistic logic', *Journal of Logic and Computation*. Online July 2013.

Expanding intuitionistic logic with a constant[‡]

Rodolfo C. Ertola Biraben*

* Centro de Lógica, Epistemologia e História da Ciência
Universidade Estadual de Campinas
rcertola@cle.unicamp.br

Foreword

It is both an honour and a pleasure to contribute to the present volume.

1 Introduction

In this paper we explore the consequences of adding a logical constant to intuitionistic logic. It is a variant of a connective that was already considered by Smetanich (1960) in the context of the so-called *intuitionistic* connectives following ideas of Novikov. Afterwards, these connectives were also considered by other authors (see e.g. Gabbay (1977) and Gabbay (1981, Chapter 7.4)). The idea behind the constant we will study is that it implies intuitionistically exactly the classical tautologies. In that way, it may be seen as a sort of local switch to classical logic. We will see that this is problematic in the case of intuitionistic predicate logic, unless we consider the extension where Glivenko's Theorem holds.

The paper is organized as follows. In Section 1 we deal with preliminary questions concerning intuitionistic logic. In Section 2 we consider the expansion of intuitionistic propositional logic with the mentioned constant. In Section 3 we provide a syntactic argument for the conservativity of the expansion. In Section 4 we consider the expansion in the case of intuitionistic predicate logic. Finally, in the last section we briefly consider the situation in the case of quantified propositional logic.

[‡]The author thanks Xavier Caicedo, who made helpful comments to an earlier version of this paper. The author has been supported by the FP7 PIRSES-GA-2009-247584 MaToMUVI Project and by the FAPESP LOGCONS Project

2 Preliminaries

Regarding notation, in what follows we will sometimes use $\mathfrak{F}$ to denote the set of formulas constructed in the usual way from an infinite set of propositional letters and a set of given connectives. Given a formula φ, the notation $lg\varphi$ will refer to the set of propositional letters in φ. The connectives $\wedge$, $\vee$, $\to$ and $\neg$ will behave as in intuitionistic propositional logic (IPL). We will use the abbreviation mp for the *modus ponens* rule and use the notation $\Gamma \vdash \varphi$ to indicate, as usual, that there is a derivation of φ from Γ in IPL. In particular, $\varphi \vdash$ and $\vdash \varphi$ mean that φ derives every formula and every formula derives φ, respectively. The letter c will refer to classical logic. Regarding terminology, we will say that we *extend* a logic when we add axioms in the language of the given logic. When we add new connectives and axioms to the language, we will speak about *expansions*. So, for instance, we will speak of conservative *expansions*, not of conservative extensions.

Firstly, let us remember the following celebrated result of Glivenko (1929), which we state without proof.

Glivenko's Theorem. $\vdash_c \varphi$ iff $\vdash \neg\neg\varphi$.

This can be easily seen to be equivalent to the following:

Corollary. $\varphi \vdash_c$ iff $\varphi \vdash$.

However, it will be useful to bear in mind that Glivenko's Theorem does not hold in the case of intuitionistic predicate logic, as it fails in the case of universally quantified formulas.

Secondly, we will use soundness with respect to valuations in Heyting algebras:

Soundness. *If $\Gamma \vdash \varphi$, then, for every Heyting algebra H and valuation v, it holds that if $v\psi = 1$, for every ψ in Γ, then $v\varphi = 1$.*

Thirdly, let us state and prove the following lemma that appears in Komori (1978, Theorem 2.1).

Lemma 1. *If $\varphi \vdash \psi$ and $lg\varphi \cap lg\psi = \emptyset$, then either $\varphi \vdash$ or $\vdash \psi$.*

Proof. Suppose i) $\varphi \nvdash$ and ii) $\nvdash \psi$. From ii) it follows that there is a Heyting algebra H and a valuation v such that $v\psi \neq 1$. From i) it follows, using Glivenko's Corollary, that $\varphi \nvdash_c$ and so, there is a valuation u in H such that $u\varphi = 1$ and, for every propositional letter p, $up = 1$ or $up = 0$. Now, as $lg\varphi \cap lg\psi = \emptyset$, we may define a valuation w in H such that $wp = up$, for every p in φ, and $wp = vp$, for every p in ψ. Then, we have a Heyting algebra H and valuation w such that $w\varphi = 1$ and $w\psi = 0$, contradicting the fact that $\varphi \vdash \psi$.   □

Remark 1. *Some logics do not enjoy the property stated in the previous lemma. Johansson's Minimal Logic m is a counterexample, as we have that $p \wedge \neg p \vdash_m \neg q$, $p \wedge \neg p \nvdash_m$, and $\nvdash_m \neg q$.*

Fourthly, let us remind the following facts, which are easy to verify:

Lemma 2. *(i)* $\vdash \neg\neg(\neg\neg\alpha \to \alpha)$.
  (ii) $\neg\neg(\alpha \wedge \beta) \dashv\vdash \neg\neg\alpha \wedge \neg\neg\beta$.
  (iii) $(\alpha \vee \neg\alpha) \wedge (\beta \vee \neg\beta) \vdash (\alpha \circ \beta) \vee \neg(\alpha \circ \beta)$, *for* $\circ \in \{\wedge, \vee \to\}$.
  (iv) $\alpha \vee \neg\alpha \vdash \neg\alpha \vee \neg\neg\alpha$.
  (v) $(\neg\neg\alpha \to \alpha) \wedge (\neg\neg\beta \to \beta) \vdash \neg\neg(\alpha \circ \beta) \to (\alpha \circ \beta)$, *for* $\circ \in \{\wedge, \to\}$.

Corollary 1. *In intuitionistic logic having $p \vee \neg p$, for every propositional letter p, implies $\varphi \vee \neg\varphi$, for every formula φ.*

However, note that in the five element Heyting algebra that results from the four element Boolean algebra adding a new top, taking the valuation v such that $v(p) = a_1$ and $v(q) = a_2$, where p and q are different propositional letters and a_1 and a_2 are the two atoms of the resulting algebra, the case $\circ = \vee$ of Lemma 2(v) does not hold. So, we have the following consequence:

Proposition 1. *Having $\neg\neg p \to p$, for every propositional letter p, does not imply, in intuitionistic logic, $\neg\neg\varphi \to \varphi$, for every formula φ.*

3 Introducing a new constant

As stated in the introduction, it is the purpose of this paper to study the consequences of expanding intuitionistic logic with a constant connective. Towards that end, consider the following fact.

Proposition 2. *There is no formula φ of IPL such that*

  (i) $\vdash_c \varphi$

and

  (ii) for all $\psi \in$ IPL, if $\vdash_c \psi$, then $\varphi \vdash \psi$.

Proof. Suppose such a formula φ exists. Let p be a propositional letter not appearing in φ. Consider the formula $p \vee \neg p$. As $\vdash_c p \vee \neg p$, using (ii) it follows that $\varphi \vdash p \vee \neg p$. But $\nvdash p \vee \neg p$. So, using Lemma 1, $\varphi \vdash$ and so, in contradicton with (i), we have $\varphi \vdash_c$.   □

Remark 2. *In the given proof, instead of using the formula $p \lor \neg\, p$, we may use $\neg\neg p \to p$.*

What happens if we add to IPL a logical constant playing the role of such a formula?

To do that, first note that (i) and (ii) in Proposition 2 are phrased in terms of classical logic. However, we want to express ourselves in intuitionistic terms. This is partly solved remembering Glivenko's Theorem. That is, instead of (i), we may as well say that $\vdash \neg\neg\varphi$ and the antecedent of the conditional in (ii) may as well be that $\vdash \neg\neg\psi$.

Also, note that it is possible to express the conditional in (ii) in the previous proposition with just one formula, as we may see now.

Proposition 3. *Let φ be a formula of IPL. Then, for any formula ψ of IPL, the following are equivalent:*

(i) *if $\vdash \neg\neg\psi$, then $\varphi \vdash \psi$,*

(ii) $\vdash \varphi \to (\neg\neg\psi \to \psi)$.

Proof. To see that (i) implies (ii), use the fact that $\vdash \neg\neg(\neg\neg\psi \to \psi)$ and the Deduction Theorem for IPL. To see that (ii) implies (i), suppose $\vdash \neg\neg\psi$, deduce from (ii) that $\varphi \vdash \neg\neg\psi \to \psi$ and conclude (i). $\square$

Note that it is easy to prove the equivalence between (ii) of Proposition 3 and (ii) of Proposition 2 without using Glivenko's Theorem.

Now we are ready to introduce a logical constant expressing the facts that $\vdash \neg\neg\varphi$ and $\vdash \varphi \to (\neg\neg\psi \to \psi)$. So, let us add to the language of IPL the logical constant γ, extend the set of formulas of IPL stating that γ is also a formula, and add to the axioms of IPL the following two axioms:

(γI) $\neg\neg\, \gamma$,

(γE) $\gamma \to (\neg\neg\varphi \to \varphi)$.

Remark 3. *Note that instead of (γI) we may use $\neg\, \gamma \to \gamma$ and instead of (γE) we may use $\gamma \to ((\neg\varphi \to \varphi) \to \varphi)$ or $\gamma \to (\varphi \lor \neg\varphi)$. Also, instead of (γE), we may use, by Corollary 1, $\gamma \to (p \lor \neg p)$, just for propositional letters p.*

Let us call IPL+γ the defined expansion of IPL.

In Smetanich (1960) (see also Rose (1962) and Yashin (1994)) the author considers a unary connective that is a variant of γ. Using our notation and the symbol $\leftrightarrow$ for the biconditional, Smetanich's unary connective would be given by the axioms $\gamma\varphi \leftrightarrow \gamma\psi$, $\neg\neg\gamma\varphi$, and $\gamma\varphi \to (\psi \lor \neg\psi)$.

Now, should γ be accepted as an *intuitionistic* connective? Regarding this issue there is a Russian tradition began by Novikov and continued by the already mentioned Smetanich and, more recently, also Yashin (see Yashin (1994) and Yashin (1998)). Not belonging to this tradition, Gabbay also has asked for certain conditions to be satisfied in order to accept a connective as intuitionistic (see Gabbay (1977) and Gabbay (1981, Chapter 7.4)). These two lines agree in requiring at least the following: (1) newness, (2) conservative expansion, and (3) the disjunction property. Gabbay also requires (4) univocity. The constant connective γ satisfies all four of them and also the finite model property and decidability at the propositional level.

Newness follows immediately from Proposition 2.

The following derivation establishes the univocity of γ:

1. γ $\qquad\qquad\qquad\qquad$ Sup
2. $\gamma \to (\neg\neg\,\gamma\,' \to \gamma\,')$ $\qquad$ γE
3. $\neg\neg\,\gamma\,' \to \gamma\,'$ $\qquad\qquad$ $1, 2, mp$
4. $\neg\neg\,\gamma\,'$ $\qquad\qquad\qquad$ $\gamma\,'I$
5. $\gamma\,'$ $\qquad\qquad\qquad\qquad$ $3, 4, mp$

In the next section we will prove that that IPL+γ is a conservative expansion of IPL. Concerning the disjunction property, in Caicedo & Cignoli (2001, p. 1632) there is an algebraic proof to the effect that the expansion of intuitionistic propositional logic with γ has that property.

Rather than introducing a logical constant playing the role of the strongest classical tautology, we may introduce a unary connective playing the role of the strongest classical tautology implied by a formula φ. This is the logical way of expressing the algebraic notion of being the smallest dense element above x, for an x belonging to the universe of a Heyting algebra, denoted as $\gamma(x)$ in Caicedo & Cignoli (2001, p. 1625). The corresponding unary connective may be given by the axioms $\neg\neg\gamma\varphi$, $\varphi \to \gamma\varphi$, and $(\varphi \to \psi) \to [\neg\neg\psi \to (\gamma\varphi \to \psi)]$, expressing that $\gamma\varphi$ is a classical tautology, that $\gamma\varphi$ is implied by φ, and that $\gamma\varphi$ is the strongest classical tautology implied by φ, respectively. Note that γ and $\gamma(\cdot)$ are interdefinable with $\gamma := \gamma\bot$ and $\gamma\varphi := \gamma \vee \varphi$.

Finally, note that γ is more present in the bibliography than what it appears to be, as it equals $S\bot$, $G\bot$, and $A\bot$, where S stands for the successor in Kuznetsov (1985) and Caicedo & Cignoli (2001), G stands for Gabbay connective in Gabbay (1977) and Gabbay (1981), and A stands for the strongest anticipator in Humberstone (2001). For more on this, see Ertola Biraben (2012).

4 IPL+γ is a conservative expansion of IPL

In order to see that IPL+γ is a conservative expansion of IPL, it is possible to give a proof using the finite model property of IPL and the following fact (for a similar situation, see Caicedo & Cignoli (2001, p. 1632):

Lemma 3. γ *exists in every finite Heyting algebra.*

Proof. If $\neg\neg y_1 = 1$ and $\neg\neg y_2 = 1$, then $\neg\neg(y_1 \wedge y_2) = 1$. $\qquad\square$

In what follows in this section, we give a constructive proof that IPL+γ is a conservative expansion of IPL. Our proof is as an adaptation of the proof in Kuznetsov (1985) for the case of the already mentioned successor connective, which can be seen as a generalization of γ, in the sense that $x \to y$ is a generalization of $\neg x$.

We first state formally that IPL+γ is a conservative expansion of IPL:

Theorem 1. *For all* $\Gamma \cup \{\alpha\} \subseteq \mathfrak{F}(IPL)$, *if* $\Gamma \vdash_{IPL+\gamma} \alpha$, *then* $\Gamma \vdash_{IPL} \alpha$.

Proof. Let $\Gamma \cup \{\alpha\} \subseteq \mathfrak{F}(IPL)$. Let us suppose that $\Gamma \vdash_{IPL+\gamma} \alpha$. Then there exists a sequence of formulas $\beta_1, \ldots, \beta_n = \alpha$ such that for all $i, 1 \le i \le n$ either (i) $\beta_i \in \text{Axioms(IPL)}$, (ii) $\beta_i \in \text{Axioms}(\gamma)$, (iii) $\beta_i \in \Gamma$, or (iv) β_i comes by mp from previous formulas in the sequence. Now, relatively to the given derivation, we define the formula

$$\delta = (p_1 \vee \neg p_1) \wedge \cdots \wedge (p_m \vee \neg p_m),$$

where the p_i are such that there is a $\beta_i = \gamma \to (p_i \vee \neg p_i)$, if there is at least one such p_i, and otherwise define $\delta = p \vee \neg p$, taking any propositional letter p.

Now, for every $i, 1 \le i \le n$, we define the formula $\epsilon_i = \beta_i[\gamma/\delta]$. Note that, then, $\epsilon_i = \beta_i$ if $\beta_i \in \Gamma$ or $i = n$.

In what follows we prove by induction that for all $i, 1 \le i \le n$, it holds that $\Gamma \vdash_{IPL} \epsilon_i$. As $\epsilon_n = \alpha$, the theorem will follow.

There are four cases to consider.

First case: $\beta_i \in \text{Axioms(IPL)}$. It follows that $\epsilon_i \in \text{Axioms(IPL)}$. Then, $\Gamma \vdash_{IPL} \epsilon_i$.

Second case: $\beta_i \in \text{Axioms}(\gamma)$. Then either $\beta_i = \neg\neg\gamma$ or $\beta_i = \gamma \to (p_i \vee \neg p_i)$. Then either $\epsilon_i = \neg\neg\delta$ or $\epsilon_i = \delta \to (p_i \vee \neg p_i)$, both of which are derivable in IPL, in the first case because of Lemma 2 (i) and (ii) and in the second case because of the axiom in IPL that states the elimination of conjunction. Then, $\Gamma \vdash_{IPL} \epsilon_i$.

Third case: $\beta_i \in \Gamma$. Then, $\epsilon_i = \beta_i$, because $\Gamma \in \mathfrak{F}(IPL)$. Then, $\Gamma \vdash_{IPL} \epsilon_i$.

Fourth case: β_i comes by mp from previous formulas $\beta_k = \beta_j \to \beta_i$ and β_j in the sequence. Here we use the induction hypothesis to get that

$\Gamma \vdash_{IPL} \epsilon_k$ and $\Gamma \vdash_{IPL} \epsilon_j$. But $\epsilon_k = \beta_k[\gamma/\delta] = (\beta_j \to \beta_i)[\gamma/\delta] = \beta_j[\gamma/\delta] \to \beta_i[\gamma/\delta] = \epsilon_j \to \epsilon_i$. Then, $\Gamma \vdash_{IPL} \epsilon_i$. $\square$

5 γ in intuitionistic predicate logic

In this section we use IQL for intuitionistic predicate logic. First of all, note that it makes sense to consider adding γ to IQL, because Proposition 2 can be generalized to the case of IQL:

Proposition 4. *There is no formula φ of IQL such that*

 (i) $\vdash_c \varphi$

and

 (ii) for all $\psi \in$ IQL, if $\vdash_c \psi$, then $\varphi \vdash \psi$.

Proof. Suppose such a formula φ exists in IQL. Let P be a unary predicate not appearing in φ, and consider the formula $\forall x(Px \lor \neg Px)$. By the interpolation property of IQL, there is a sentence σ in the language of pure identity such that $\varphi \vdash_{IQL} \sigma \vdash_{IQL} \forall x(Px \lor \neg Px)$. Since $\vdash_c \varphi$, then $\vdash_c \sigma$. Now, any Kripke model with injective transitions satisfies $\forall x \forall y(x = y \lor \neg x = y)$ and thus all classically valid formulas in the pure language of identity. Consider a model which consists of a chain where the same infinite universe is repeated countably many times and the interpretation of the predicate P is strictly increasing, then it satisfies σ but not $\forall x(Px \lor \neg Px)$, contradicting that $\vdash_{IQL} \forall x(Px \lor \neg Px)$. $\square$

Now, when we introduced γ in the context of IPL, we used Glivenko's Theorem to axiomatize it. However, in the context of IQL, Glivenko's Theorem does not hold. Accordingly, in the context of IQL, the given axioms for γ should be expected to have a different meaning. In fact, γI purports to say that γ is derivable in classical logic. This is achieved, using Glivenko's Theorem, stating $\neg\neg\gamma$ as axiom. However, in the context of IQL this turns to be stronger, as we have that if the double negation of a formula is derivable in IQL, then the formula itself is derivable in predicate classical logic, but the converse is *not* the case. Accordingly, γI might have unexpected consequences. In fact, we have the following result:

Theorem 2. $\vdash_{IQL+\gamma} \neg\neg\forall x(Px \lor \neg Px)$.

Proof. Using γE we have that $\gamma \vdash_{IQL+\gamma} \forall x(Px \lor \neg Px)$. So, using a property of negation in IPL, it follows that $\neg\neg\gamma \vdash_{IQL+\gamma} \neg\neg\forall x(Px \lor \neg Px)$. Now, using γI, it follows that $\vdash_{IQL+\gamma} \neg\neg\forall x(Px \lor \neg Px)$. $\square$

The problem is that the formula $\neg\neg\forall x(Px \vee \neg Px)$ belongs to the language of IQL, and is very well known not to be derivable in IQL. So, IQL+γ is *not* a conservative expansion of IQL. The given theorem also solves a case of a problem raised in López-Escobar (1985, p. 118), that is, to find a conservative expansion of IPL that derives in IQL the schema corresponding to the formula in Theorem 2.

It is equivalent to extend IQL with any of the schemas generalizing the formula $\neg\neg\forall x(Px \vee \neg Px)$ or $\forall x\neg\neg Px \rightarrow \neg\neg\forall xPx$, or Glivenko's Theorem. These equivalences already appear in Gabbay (1972).

In absence of Glivenko's Theorem, it seems that it is not possible to express with a logical constant in IQL the existence of a formula satisfying the conditions of Proposition 4.

Now, let us note the following. We have just seen that, according to Theorem 2, in IQL+γ we may derive the formula $\neg\neg\forall x(Px \vee \neg Px)$, which is not derivable in IQL. Let us call MH to the extension of IQL with any of the given schemas that ensure Glivenko's Theorem. It happens that in MH the connective γ is not definable. In order to see this, note that MH has the interpolation property, which was proved semantically in Gabbay (1971) and proof-theoretically in Seldin (1986). Then we can apply the same argument as in the case of Proposition 4.

The situation is somewhat different in the case of complete Heyting algebras. To see this, let us first note that, from an algebraic point of view, γ may be defined as the $min\{x : \neg\neg x = 1\}$. It follows that

(C) $\neg\neg x = \gamma \rightarrow x$.

Now, we have the following fact:

Theorem 3. *Let H be a complete Heyting algebra. Then γ exists if and only if, for every family $\{y_i\}_{i\in I}$, we have that $\bigwedge_i \neg\neg y_i = \neg\neg \bigwedge_i y_i$.*

Proof. $\Rightarrow$) Suppose that γ exists. The inequality $\neg\neg \bigwedge y_i \leq \bigwedge \neg\neg y_i$ follows by monotonicity of $\neg\neg$ without the need of the existence of γ. In order to see the other inequality, use (C) twice. It is enough to see that $\bigwedge(\gamma \rightarrow y_i) \leq \gamma \rightarrow \bigwedge y_i$. Indeed, $\bigwedge(\gamma \rightarrow y_i) \wedge \gamma \leq \bigwedge(\gamma \wedge (\gamma \rightarrow y_i)) = \bigwedge(\gamma \wedge y_i) = \gamma \wedge \bigwedge y_i \leq \bigwedge y_i$.

$\Leftarrow$) Reciprocally, suppose that the equality holds. It is enough to see that $\bigwedge\{t : \neg\neg t = 1\}$ belongs to $\{t : \neg\neg t = 1\}$. Indeed, $\neg\neg(\bigwedge_{\neg\neg t=1} t) = \bigwedge_{\neg\neg t=1} \neg\neg t = 1$. $\square$

Finally, it is clear that from the theorem just proved we may deduce that the expansion of MH with γ is conservative.

6   Adding propositional quantification

In this section we consider the behaviour of γ in the context of propositional quantification, i.e. intuitionistic logic expanded with the universal and the existential propositional quantifiers, which we shall call IQPL. Regarding this logic, the reader may consult e.g. Gabbay (1977, Chapter 9).

Similarly to the case of IQL+γ, the expansion of IQPL with γ is not conservative, as the formula $\neg\neg\forall p(p \vee \neg p)$ is derivable. However, in the case of IQPL there exists a formula satisfying the conditions of Proposition 2 or 4, to wit, the formula $\forall p(p \vee \neg p)$.

We may also also consider the expansion of IQPL obtained by adding to the language a constant connective δ satisfying the following two axioms:

(δI) $\exists p(\delta \leftrightarrow (\neg p \vee \neg\neg p))$,
(γE) $\delta \rightarrow (\neg\varphi \vee \neg\neg\varphi)$.

In this case we use the existential quantifier in the introduction axiom, as we do not know whether there exists a formula that plays the role played by $\neg\neg\gamma$ in the case of the constant γ. Note, also, that in the case of γ we may use the formula $\exists q(\gamma \leftrightarrow (q \vee \neg q))$ instead of $\neg\neg\gamma$. This was already noticed for any formula instead of γ in Kreisel (1981, p. 10) and is easy to prove. It corresponds to the algebraic fact that $\neg\neg x = 1$ is equivalent to the existence of a y such that $x = y \vee \neg y$.

Similarly to the case of γ, the expansion with δ will not be conservative, as it will be possible to derive the formula $\neg\neg\forall p(\neg p \vee \neg\neg p)$.

However, it would be more appropriate to say that we can prove the following proposition, where we call dM the extension of IQPL with the schema $\neg\neg\varphi \vee \neg\varphi$:

Proposition 5. *There is a formula φ of IQPL such that*

   (i) $\vdash_{dM} \varphi$

and

   (ii) for all $\psi \in$ IQPL, if $\vdash_{dM} \psi$, then $\varphi \vdash_{IQPL} \psi$.

Proof. The formula $\forall p(\neg\neg p \vee \neg p)$ fulfills both conditions. $\square$

This way of proceeding can be generalized.

References

Caicedo, X. & Cignoli, R. (2001), 'An algebraic approach to intuitionistic connectives', *The Journal of Symbolic Logic* 66(4), 1620–1636.

Ertola Biraben, R. C. (2012), 'On some extensions of intuitionistic logic', *Bulletin of the Section of Logic* **41**(1/2), 17–22.

Gabbay, D. M. (1971), Semantic proof of Craig's interpolation theorem for intuitionistic logic and extensions. part ii, *in* R. O. Gandy & C. M. E. Yates, eds, 'Logic Colloquium '69', North-Holland, Amsterdam, pp. 403–410.

Gabbay, D. M. (1972), 'Applications of trees to intermediate logics', *The Journal of Symbolic Logic* **37**, 135–138.

Gabbay, D. M. (1977), 'On some new intuitionistic propositional connectives. I', *Studia Logica* **36**, 127–139.

Gabbay, D. M. (1981), *Semantical investigations in Heyting's intuitionistic logic*, Vol. 148, Springer.

Glivenko, V. I. (1929), 'Sur quelques points de la Logique de M. Brouwer', *Bulletin de la Classe des Sciences de l'Académie Royale de Belgique* **15**, 183–188.

Humberstone, L. (2001), 'The pleasures of anticipation: enriching intuitionistic logic', *Journal of Philosophical Logic* **30**, 395–438.

Komori, Y. (1978), 'Logics without Craig's interpolation property', *Proc. of the Japan Academy* **54**, 46–48.

Kreisel, G. (1981), 'Monadic operators defined by means of propositional quantification in intuitionistic logic', *Reports on Mathematical Logic* **12**, 9–15.

Kuznetsov, A. V. (1985), 'On the propositional calculus of intuitionistic provability', *Soviet Math. Dokl.* **32**(1), 18–21.

López-Escobar, E. G. K. (1985), 'On intuitionistic connectives I', *Revista Colombiana de Matemáticas* **XIX**, 117–130.

Rose, G. F. (1962), 'A680', *Mathematical Reviews* **24**, 129–130.

Seldin, J. P. (1986), 'On the proof theory of the intermediate logic MH', *The Journal of Symbolic Logic* **51**(3), 626–647.

Smetanich, Y. S. (1960), 'On the completeness of a propositional calculus with a supplementary operation in one variable', *Tr. Mosk. Mat. Obsch.* **9**, 357–371.

Yashin, A. D. (1994), 'The Smetanich logic T and two definitions of a new intuitionistic connective', *Mathematical Notes* **56**, 745–750.

Yashin, A. D. (1998), 'New solutions to Novikov's problem for intuitionistic connectives', *Journal of Logic and Computation* **8**(5), 637–664.

Bounded dialectica interpretation: categorically[‡]

Valeria de Paiva*

* Nuance Communications
1198 E. Arques Ave, Sunnyvale
valeria.depaiva@nuance.com

Abstract

Ferreira and Oliva introduced the Bounded Functional Interpretation (BFI) in Ferreira & Oliva (2005) as a way of continuing Kohlenbach's programme of shifting attention from the obtaining of precise witnesses to the obtaining of *bounds* for these witnesses, when proof mining. One of the main advantages of working with bounds, as opposed to witnesses, is that the non-computable mathematical objects whose existence is claimed by various ineffective principles can sometimes be bounded by computable ones. In this note we present first steps towards a categorical version of BFI, along the lines of de Paiva's version of Gödel's Dialectica interpretation, the Dialectica constructions, in de Paiva (1990). The previous categorical constructions seem to extend smoothly to the new ordered setting.

1 Introduction

This preliminary note tries to make good in the promise I have made over the years to Luiz Carlos Pereira to connect the categorical constructions of my thesis, the Dialectica Categories, de Paiva (1990), to the actual proof theory that inspired them, that is, to the functional interpretations themselves, mostly Gödel's Dialectica interpretation, but also its Diller and Nahm variant.

This note is much indebted to the careful work done by Paulo Oliva and collaborators, especially Gilda Ferreira, Ferreira & Oliva (2009), as I hope it is abundantly clear from the note itself. But, not surprisingly, I disagree with some of their conclusions, hence the need to write this. Lastly, a word of caution: it is possible that all that I have to say here is better said

---

[‡]This paper is dedicated to Luiz Carlos Pereira on his birthday.

(and has been said) in the higher-order (in the sense of category theory) language of Hofstra (2011) and Hyland (2002). But there is still a point in writing this, as the "translation" from the fibrations and higher-order category theory language to the pedestrian category theory used here is non trivial and many will not be able to read the very abstract formulation.

2 The Dialectica constructions

For my thesis I was originally trying to provide an internal categorical model of Gödel's Dialectica Interpretation, which I presumed would be a cartesian closed category. But the categories I came up with proved also to be models of Linear Logic. This was a surprise and somewhat of a boost for Linear Logic, which was only beginning to appear then.

The traditional categorical modeling of intuitionistic logic goes as follows: a formula A is mapped to an object A of an appropriate category, the conjunction $A \wedge B$ is mapped to the cartesian product $A \times B$ and the implication $A \to B$ is mapped to the space of functions B^A (the set of functions from A to B). These are real cartesian products, so we have projections ($A \times B \to A$ and $A \times B \to B$) and diagonals ($A \to A \times A$), which correspond to deletion and duplication of resources. This is not a linear structure. To model a linear logical system faithfully we need to use *tensor products* and *internal homs* in Category Theory. Luckily these structures were considered by category theorists long before Linear Logic, so they were easy. What was hard was to define the "make-everything-usual" operator, the modality !, which applied to a linear proposition, makes it an intuitionistic one.

Definition 1 *The category* $\text{Dial}_2(\textbf{Sets})$ *has as objects triples* $A = (U, X, \alpha)$, *where* U, X *are sets and* α *is an ordinary relation between* U *and* X. *(so either* u *and* x *are* α *related,* $\alpha(u, x) = 1$ *or not.)*

A map from $A = (U, X, \alpha)$ *to* $B = (V, Y, \beta)$ *is a pair of functions* (f, F), *where* $f \colon U \to V$ *and* $F \colon X \to Y$ *such that*

$$\forall u \in U, \forall y \in Y \quad \alpha(u, Fy) \text{ implies } \beta(fu, y)$$

or $\alpha(u, F(y)) \leq \beta(f(u), y)$

An object A is not symmetric: think of (U, X, α) as $\exists u. \forall x. \alpha(u, x)$, a proposition in the image of the Dialectica interpretation. But using this category and its linear, multiplicative structure we can prove:

Theorem 1 (Model of CLL modality-free, 1988) *The category* Dial_2 *(Sets) has products, coproducts, tensor products, a par connective, units for the four monoidal structures and a linear function space, satisfying the appropriate categorical properties for these structures. The category* Dia_2*(Sets) is a symmetric monoidal closed category with an involution* $(\ )^*$ *that makes it a model of classical linear logic, without exponentials.*

But how do we get modalities, or, as Girard calls them, exponentials? For this specific categorical model, we need to define two special co-monads and do plenty of work using distributive laws to prove the desired theorem. Recall that a 'modalized' object $!A$ must satisfy $!A \to !A \otimes !A$, $!A \otimes B \to I \otimes B$, $!A \to A$ and $!A \to !!A$, together with several equations relating them. The difficulty is to define a comonad "!" such that its coalgebras are commutative comonoids and such that the coalgebra and the comonoid structures interact nicely.

Theorem 2 (Model of Classical Linear Logic with modalities, 1988) *We can define comonads T and S on* Dial_2*(Sets) such that the Kleisli category of their composite* $! = T; S$, Dial_2*(Sets)$^!$ is cartesian closed.*

Define T by saying $A = (U, X, \alpha)$ goes to (U, X^*, α^*) where X^* is the free commutative monoid on X and α^* is the multiset version of α.

Define S by saying $A = (U, X, \alpha)$ goes to (U, X^U, α^U) where X^U is the set of functions from U into X. Then compose T and S to get $A = (U, X, \alpha)$ goes to $(U, (X^*)^U, \alpha*^U)$. This composite comonad does get us from the linear category to the cartesian closed category corresponding to the Diller-Nahm interpretation. This was the ultimate result desired in the thesis. But on the way to proving this we had a second main definition

Definition 2 *The category* DDial_2*(Sets) has as objects triples $A = (U, X, \alpha)$, where U, X are sets and α is an ordinary relation between U and X.*

A map from $A = (U, X, \alpha)$ to $B = (V, Y, \beta)$ is a pair of functions (f, F), where $f: U \to V$ and $F: U \times X \to Y$ such that

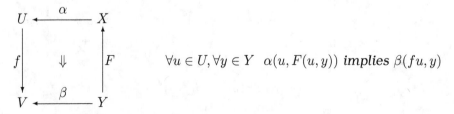

$\forall u \in U, \forall y \in Y \ \ \alpha(u, F(u, y))$ *implies* $\beta(fu, y)$

$$\text{or } \alpha(u, F(u,y)) \leq \beta(f(u), y)$$

This category seems extremely similar to our first definition, in particular the objects of the two categories are exactly the same. But morphisms are what make a category and morphisms in the two cases are different enough, and induce very different structures in the categories. In particular for this category we cannot define a multiplicative disjunction, unlike the previous one, and the tensor product for the second category is much simpler.

Theorem 3 (Model of ILL modality-free, 1987) *The category* DDial$_2$ (Sets) *has products, (weak)-coproducts, tensor products, units for the two monoidal structures and a linear function space, satisfying the appropriate categorical properties. The category* DDial$_2$(Sets) *is a symmetric monoidal closed category with products and weak coproducts, which makes it a model of* **intuitionistic** *linear logic, without exponentials.*

This model of restricted, intuitionistic only linear logic has nonetheless a very special property: it has *co-free* comonoids and hence it's a rather special model of linear logic, which accounts for the theorem below:

Theorem 4 (Model of Intuitonistic Linear Logic with modalities, 1987) *We can define a co-free comonad* $!_{DN}$ *on* DDial$_2$(Sets) *such that its co-Kleisli category* Dial$_2$(Sets)$^!$ *is cartesian closed.*

Many calculations are needed to prove that the linear logic modality ! is well-defined and to obtain a model of *classical* linear logic eventually. But my previous work did not emphasize this two-step process, as the goal was to obtain a cartesian closed category, a model of intuitionistic linear logic. The two step process goes from Dial to DDial, from DDial to DNDial, the Diller-Nahm category, which is now Cartesian Closed. Here we want to see this two-step process as a reflexion of what is happening in the proof theory. The point being that going from the linear logic category Dial to the intuitonistic linear logic category DDial, which has the same objects, but different morphisms and hence different function spaces and different tensors can be seen as the result of applying the modified realizability translation, in Oliva's terms. These we recap briefly below.

3 Unification of Proof Interpretations

A recent preprint Oliva (2014) provides a summary of Oliva's proposed syntactic unification of functional interpretations, a long standing research program that started with the unification of Gödel's Dialectica Interpretation and Kreisel's Modified Realizability in Oliva (2006). This unification is purely *syntactic*, so formulae of logical systems are mapped to other logical systems, but proofs of propositions do not have to be mapped, a priori.

In principle a semantic version of this unification already exists for the Dialectica interpretation and its Diller-Nahm variant, as result of the work in de Paiva (1990). The basis of Oliva's more comprehensive syntactic unification (which includes besides Diller-Nahm and Dialectica, modified realizability and their "truth"-variants, five functional interpretations so far) is the rephrasing of modified realizability in terms of *relational realizability* which talks about a relation between potential witnesses and challenges, in a way very similar to the way the dialectica interpretation is usually presented in terms of witnesses and counter-examples.

A second main idea of this syntactic unification is described in Oliva (2010); Ferreira & Oliva (2009) where it is proposed that the main characteristics of the functional interpretations can be gotten via interpretation of intuitionistic linear logic, complemented by several different modalities, in particular modalities for Gödel's dialectica, for the Diller-Nahm variant of dialectica and for modified realizability.

While Oliva's program acknowledges previous work of the present author de Paiva (1989b,a) as inspiration, the exact categorical relationships are not clearly spelled out. In particular the categorical modelling of Kreisel's modified realizability, in terms of realizability toposes and the work developed in a series of papers by van Oosten, Hyland and others is not, as yet, related to this unification. The goal of this note is to see where are the problems in making this connection and whether we can see a way of solving them. We start by reviewing the goals and motivations of both realizability and functional interpretations.

3.1 Realizability

Kleene (1973) recounts how Kleene's idea for a numerical realizability developed as he wished to give some precise meaning to the intuition that there should be a connection between Intuitionism and the theory of recursive functions. Both theories stress the importance of extracting information effectively. Kleene starts by conjecturing a weak form of Church's Rule: if a closed formula of the form $\exists x \forall y \phi(x, y)$ is provable in intuitionistic number theory, then there must exist a general recursive function F such that for all n, the formula $\phi(n, F(n))$ is true.

Kleene's main motivation for inventing realizability in the forties seems to be the work of Hilbert and Bernays, in their Grundlagen der Mathematik. They explained the finitist position in Mathematics, as follows (translation by Jaap van Oosten in Van Oosten (2002))

> An existential statement about numbers, i.e. a statement of the form 'there exists a number n with property $A(n)$' is finitistically taken as a 'partial judgement' that is, as an incomplete rendering of a more precisely determined proposition which consists in either giving directly a number n with the property $A(n)$ or a procedure by which such a number can be found...

Kleene wondered whether a completion procedure might be provided that completed the description for all logical connectives, but suggested that his was, in any case, a "partial analysis of the intuitionistic meaning of the statements".

Realizability is famously understood in categorical terms using Hyland's "Effective topos" Hyland (1982). Modified realizability, as originally defined by Kreisel, has also given rise to a categorical formulation in terms of toposes, so one would expect connections between these toposes and Oliva's program of functional interpretations unification, at the semantic level.

3.2   Proof Mining and Functional Interpretations

Proof mining is the process of logically analyzing proofs in mathematics with the aim of obtaining new information. How do we do proof mining and what do we get? A proof of a theorem like "x as an element of a space X is a root of a function $f : X \to R$" is a complete theorem, i.e. it gives us an equation $f(x) = 0$ and we don't need any other information. But a theorem stating that "f is (strictly) positive at a point $x \in X$" is incomplete, for it leaves open how far from zero the value $f(x)$ actually is. It turns out that in many cases the information missing in an incomplete theorem can be extracted by purely logical analysis out of prima-facie ineffective proofs of the theorem. The *Dialectica interpretation* together with other proof interpretations is a major tool for proof mining. But what are proof interpretations or functional interpretations? They are proof transformations that can be used to extract extra information from proofs. Ulrich Kohlenbach has made extensive use of these and explains a full range of proof interpretations and their applications in Kohlenbach (2008).

4   Semantic Unification of Proof Interpretations?

Most of the recent work on semantic proof interpretations has been using fibrations, e.g. Hyland (2002), Biering (2008), Hofstra (2011) and Hedges (2014). While this more expressive framework might be necessary, the syntactic unification described by Oliva and collaborators gives hope that another, more pedestrian route might be available. And if the pedestrian route is not available, it will be enlightening to see where it fails vis-à-vis the syntactic work.

Taking a leaf from Shirahata (2006) and Oliva and Ferreira's work we should discuss the proof interpretation of a weaker system first. While my original work was mostly concerned with intuitionistic logic, and considered Linear Logic as a step stone to get to "traditional" logic, some of the newer work, starting with Shirahata, is about interpretations of *linear* logical systems. Some are about classical linear logic, some about intuitionistic linear or affine logic.

4.1   Intuitionistic Linear Logic

We revisit "Functional Interpretation of Intuitionistic Linear Logic", Ferreira & Oliva (2009), where we disagree, somewhat, with their interpretation of the connection between our works. They say and we agree that

> In this section we try to explain and make more explicit the link
> between our framework for unifying interpretations of IL via
> interpretations of ILL and the categorical approach on de Paiva
> (1990, 1989a,b) for modelling ILL.

But while it is true that "in de Paiva (1989a) one finds a categorical version of the Dialectica interpretation and an endofunctor interpretation for the modality !A that corresponds to the Diller-Nahm interpretation" the table setting up the correspondence in their paper is not as precise as it needs to be. In particular, what is missing from the explanation in section 5 of Ferreira & Oliva (2009) is the two-step process that takes us from modality-free Linear Logic to Intuitionistic Logic. The first of these steps sends us from a pure (no modalities) linear category to one that might correspond to Modified Realizability in their terms, while the second step takes us from modified realizability objects to Diller-Nahm sets of witnesses/challenges.

Note that the realisers of the functional interpretation are taken from a given (fixed) cartesian closed category in de Paiva (1990), while Ferreira and Oliva work with the particular cartesian closed category of the functionals of finite type. Apart from the relation $\alpha \oplus \beta$, the interpretation of the

Table 1: Comparison

	Fer-Oliva	DDial	Dial
realizers	finite types	ccc C	ccc C
formulas	$\|A\| \subseteq U \times X$	$\alpha \subseteq U \times X$	$\alpha \subseteq U \times X$
sequents	$A \vdash B$	$(f : U \to V, F : U \times Y \to X)$	$(f : U \to V, F : Y \to X)$
linear impl	$A \multimap B$	$(V^U \times X^{U \times Y}, U \times Y, \alpha \multimap \beta)$	$(V^U \times X^Y, U \times Y, \alpha \multimap \beta)$
tensor	$\otimes$	$(U \times V, X \times Y)$	$(U \times V, X^V \times Y^U)$
modality $!_R$	$!\forall x \|A\|_x^u$		(U, X^U, α^U)
modality $!_{DN}$	$!\forall x \in a \|A\|_x^u$	(U, X^*, α^*)	
modality $!_{R;DN}$	$!\|A\|_x^a$		$(U, (X^U)^*, (\alpha^U)^*)$

linear logic connectives in DDial coincides precisely with the definitions in Ferreira & Oliva (2009).

Note that Ferreira & Oliva (2009) assumes that each finite type is inhabited by at least one element, while de Paiva (1990) imposes no similar restriction. (This means that Ferreira and Oliva are dealing with affine logic, instead of linear logic, as the tensor operator can now have projections given by the inhabitants of the sets.) In the category DDial we have no way of producing projection functions $\pi_{1,2} : A \times B \to A, B$, so we are dealing with linear logic, not affine logic.

While we do explain (in page 39 of de Paiva (1990)) how one could, in principle 'cheat' and provide an interpretation for the contraction axiom $A \multimap A \otimes A$ by using a trick to map into one counterexample, when we have a choice of two such, this is not a uniform operation. Hence it does not produce a functor, is poor category theory and we do not pursue it in our work.

Another difference is the weak coproduct of de Paiva (1990), which can be made much simpler in Ferreira and Oliva's setting using inhabitedness of witnesses and counter-examples. In the categorical approach we do not assume inhabitedness, but can still obtain a generic weak coproduct.

4.2   Modified Realizability Modality

If one thinks of modified realizability witnesses as actual and potential witnesses, as in Oliva's reformulation, where actual witnesses are a subset of the potential ones, then objects of the form (U, X^U, α^U) make some intuitive sense, as we can think of U as being naturally embedded into X^U via the "constant" map $X \to X^U$. Then we might end up with a construction as follows:

Definition 3 *Objects of the category* MR *(for modified realizability) are triples $A = (U, X^U, \alpha^U)$, $B = (V, Y^V, \beta^V)$ where the relation $\alpha^U : U \times X^U \to 2$ is the composition of $U \times X^U \to U \times X \to^\alpha 2$. Morphisms need to be* considered in the category of coalgebras of the comonad R that sends an object A to $RA = (U, X^U, \alpha^U)$

Ferreira and Oliva state, when discussing the bang interpretation that "Our conditions are more general, and include as particular case the instance where "!" is a comonad with comonoid objects." It is true that their conditions are more general, but they are simply of the form "this needs to happen", while the categorical model exhibits a mathematical structure that indeed satisfies the requirements of the model.

4.3   Diller-Nahm Modality

If we start from the category Dial we can simply apply the Diller-Nahm modality taking (U, X, α) to (U, X^*, α^*) where we are considering not simply the free monoid in X, but actually the free *commutative* monoid on X. This corresponds precisely to the idea that for the Diller-Nahm interpretation we collect the witnesses into a finite, but unlimited set.

4.4   The composite $!_{R,DN}$ Modality

Only when we compose the two comonads we can get that objects of the form $!A$ satisfy all the conditions for modelling propositions of intuitionistic logic.

5   Dialectica Spaces over Partial Orders

Let us consider the category Poset of partially ordered sets and monotone functions. Monotone functions compose to give monotone functions and the identity function on a poset $(X, \leq)$ is monotonic. The category Poset has products. Given $(X, \leq_X)$ and $(Y, \leq_Y)$ their product is $(X \times Y, \leq_{X \times Y})$, where the order on the product is pointwise order $(x, y) \leq_{X \times Y} (x', y')$ iff $x \leq_X x'$ and $y \leq_Y y'$.

The category Poset is cartesian closed, the function space of Poset is given by monotone maps ordered pointwise $f \leq g : U \to V$ iff $f(u) \leq_V g(u)$ for all u in U. The category Poset also has coproducts, given by the disjoint sum.

What can we say about the dialectica constructions over the category Poset? In previous work we described two Dialectica-constructions: they share the objects, but morphisms in DDial are more complicated than the ones in Dial. The category Dial was originally called GC as a thank-you note to Girard, who originally suggested it as a simplification of the dialectica DDial construction.

First we can define the category $\text{Dial}_2\text{Posets}$.

Definition 4 *The objects of the category $\text{Dial}_2\text{Posets}$ are triples $A = (U, X, \alpha)$ where $(U, \leq_U)$ and $(X, \leq_X)$ are posets and $\alpha\colon U \times X \to 2$ is a generalized relation into 2. Maps are monotone maps $f\colon U \to V$ and $F\colon Y \to X$ such that $\alpha(u, F(y)) \leq \beta(f(u), y)$. This means that in the following commuting diagram we have a 2-cell:*

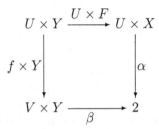

The identity map on an object $A = (U, X, \alpha)$ consists of the identity on U and the identity in X, (id_U, id_X) which clearly satisfies the implication condition.

Composition of morphisms is straightforward. Given morphisms $(f, F)\colon A \to B$, where B is (V, Y, β) and $(g, G)\colon B \to C$, where C is (W, Z, γ), the composition in the first coordinate is simply $f; g\colon U \to W$ and in the second coordinate we have $Z \xrightarrow{G} V \xrightarrow{F} X$. We are always composing monotone maps, which gives us monotone maps and we need to check that $\alpha(u, F(G(z))) \leq \gamma(g(f(u)), z)$. But because (f, F) is a morphism we know $\alpha(u, F(G(z))) \leq \beta(f(u), G(z))$ and because (g, G) is a morphism we know $\beta(f(u), G(z)) \leq \gamma(g(f(u)), z)$, so putting the inequalities together we have the desired one.

As discussed in de Paiva (2007) we need proper products and function spaces in Posets to define tensor products and function spaces in $\text{Dial}_2\text{Posets}$. Since the category Posets does have products and function spaces we can define

Definition 5 *Given objects A and B, say (U, X, α) and (V, Y, β), of Dial_2Posets their function space $A \multimap B$ is the object $(V^U \times X^Y, U \times Y, \alpha \multimap \beta)$ where the relation $\alpha \multimap \beta$ is defined by $\alpha \multimap \beta([h, H], [u, y])$ iff whenever $\alpha(u, H(y))$ holds then $\beta(h(u), y)$ holds.*

This is a direct internalization of the notion of dialectic morphism above and reverse engineering from this notion of morphism (to produce a monoidal closed category) gives us the following notion of tensor product.

Definition 6 *Given objects A and B of $\text{Dial}_2\text{Posets}$ $((U, X, \alpha)$ and (V, Y, β) respectively), their tensor product $A \otimes B$ is the object $(U \times V, X^V \times Y^U, \alpha \otimes \beta)$ where the relation $\alpha \otimes \beta$ is defined by $\alpha \otimes \beta([u, v], [h_1, h_2])$ iff whenever $\alpha(u, h_1(v))$ and $\beta(h(v, h_2(v))$ hold.*

Then, all going according to plan, we end up with a theorem that says:

Theorem 5 *The category $\text{Dial}_2\text{Posets}$ is a symmetric monoidal closed category, with products and coproducts.*

Products are given by products in the first coordinate and coproducts in the second, so if $A = (U, X, \alpha)$ and $B = (V, Y, \beta)$ then the product $A \times B = (U \times V, X + Y, p)$ where $p : U \times V \times (X + Y) \to 2$ chooses either α or β depending on the element of $X + Y$ picked. A terminal object will be $1 = (1, 0, \iota)$ where ι is the empty relation on 1×0. Note that the previous construction of modalities has not even been attempted here, so far.

Natural Numbers Objects

The category of sets has the original natural numbers object N, with zero and successor zero, suc functions. One might wonder about a natural numbers object in Dial_2Sets. By analogy one would expect it to be something like (N, X, α), where N is the natural numbers in sets and X would be some kind of dual of natural numbers. While a dual of the natural numbers is hard to conceive in sets, it would make sense to simply invert the order, if considering the concept in Poset.

Say we had such a natural numbers object $N = (N, N, \alpha)$ in D_2Posets, where $\alpha : N \times N \to 2$ is the diagonal. As discussed in De Paiva et al. (2014), were we to consider a NNO structure with respect to the categorical product in Dial_2Sets, we would need a map zero $z : 1 \to N$. Since the terminal object in Dial_2Sets is $(1, 0, ch)$, the empty relation in the empty set, this would be $(z : 1 \to N, \overline{z} : N \to 0)$ and there is no map $\overline{z} : N \to 0$. But degenerate linear NNO's, using the monoidal structure of the tensor product in Dial_2Sets can be obtained as suggested in the note above.

6 Conclusions

We have re-appraised the work on Dialectica constructions under the light of the unification of functional interpretations carried out by Gilda Ferreira and Paulo Oliva Ferreira & Oliva (2009). The main result is our improved table of comparisons between our works.

Further we described an easy extension of the formalism of Dialectica constructions to ordered categories, which we claim should model the bounds on witnesses and counter-examples as described in the Bounded Functional Interpretation of Fernando Ferreira and Paulo Oliva Ferreira & Oliva (2005). We have not solved the main problem we wished to investigate, that is the connection between the Effective Topos (and the Modified Realizability topos) and the categorical working arising from Dialectica and its connection to Linear Logic. But we have already 'borrowed' more time than we can afford, to finish this note. It is a small souvenir (*uma lembrancinha*) for Luiz Carlos and it will have to do.

References

Biering, B. (2008), Dialectica Interpretations, a Categorical Analysis, PhD thesis, PhD thesis, IT University of Copenhagen.

de Paiva, V. (1989a), 'The dialectica categories', *Categories in Computer Science and Logic* 92, 47–62.

de Paiva, V. (1990), The dialectica categories, PhD thesis, University of Cambridge, Computer Lab.

de Paiva, V. (2007), 'Dialectica and chu constructions: Cousins?', *Theory and Applications of Categories* 17(7), 127–152.

de Paiva, V. C. (1989b), A dialectica-like model of linear logic, *in* 'Category Theory and Computer Science', Springer Berlin Heidelberg, pp. 341–356.

De Paiva, V., Morgan, C. & Da Silva, S. G. (2014), 'Natural number objects in dialectica categories', *Electronic Notes in Theoretical Computer Science* 305, 53–65.

Ferreira, F. & Oliva, P. (2005), 'Bounded functional interpretation', *Annals of Pure and Applied Logic* 135(1), 73–112.

Ferreira, G. & Oliva, P. (2009), Functional interpretations of intuitionistic linear logic, *in* 'Computer Science Logic', Springer, pp. 3–19.

Hedges, J. (2014), 'Dialectica models of additive-free linear logic', *CoRR*. URL http://arxiv.org/abs/1401.4538.

Hofstra, P. (2011), The dialectica monad and its cousins, *in* C. Proceedings & L. N. . 107-139, eds, 'Models, Logics, and Higher-dimensional Categories: A Tribute to the Work of Mihaly Makkai'.

Hyland, J. (2002), 'Proof theory in the abstract', *Annals of Pure and Applied Logic* **114**(1), 43–78.

Hyland, J. M. E. (1982), 'The effective topos', *Studies in Logic and the Foundations of Mathematics* **110**, 165–216.

Kleene, S. (1973), Realizability: a retrospective survey, in 'Cambridge summer school in mathematical logic', Springer, pp. 95–112.

Kohlenbach, U. (2008), *Applied proof theory: proof interpretations and their use in mathematics*, Springer.

Oliva, P. (2006), 'Unifying functional interpretations', *Notre Dame Journal of Formal Logic* **47**(2), 263–290.

Oliva, P. (2010), 'Functional interpretations of linear and intuitionistic logic', *Information and Computation* **208**(5), 565–577.

Oliva, P. (2014), 'Unifying functional interpretations: Past and future', *arXiv preprint arXiv:1410.4364*.

Shirahata, M. (2006), 'The dialectica interpretation of first-order classical affine logic', *Theory and Applications of Categories* **17**(4), 49–79.

Van Oosten, J. (2002), 'Realizability: a historical essay', *Mathematical Structures in Computer Science* **12**(03), 239–263.

Contradictions, inconsistencies and other oxymora

Walter Carnielli*, Abilio Rodrigues[†]

* CLE and Department of Philosophy
State University of Campinas
walter.carnielli@cle.unicamp.br

[†] Department of Philosophy
Federal University of Minas Gerais
abilio@ufmg.br

Abstract

The aim of this paper is twofold. Firstly, we aim at presenting and defending a view of the nature of contradictions that intends to provide a philosophical justification for paraconsistent logics, in particular for the logics of formal inconsistency. The view we shall present here is not committed to dialetheism, the thesis according to which there are true contradictions. The second purpose consists of presenting the outline of an analysis of three concepts central to paraconsistent logics in general, namely, contradiction, negation and consistency.

1   Introduction

The aim of this paper is twofold. Firstly, we aim at presenting and defending a view of the nature of contradictions that intends to provide a philosophical justification for paraconsistent logics, in particular for the logics of formal inconsistency (*LFIs*), a family of paraconsistent logics that encompasses the great majority of paraconsistent systems developed within the Brazilian and Polish traditions, although not restricted to them (cf. Carnielli & Marcos, 2001; Carnielli et al., 2007). The view we shall present here is not committed to dialetheism, the thesis according to which there are true contradictions. Contradictions are conceived here from an epistemic viewpoint, they cannot obtain ontologically. We want to make clear that the philosophical significance of the logics of formal inconsistency is independent of one's beliefs regarding 'real contradictions'.

Our second purpose, sympathetically provocative, is an attempt to show that the concepts of contradiction, negation, and consistency, central to paraconsistent logics in general, are much more complex than the standard (i.e. classical) understanding of them suggests. With respect to the concept of consistency, we will argue that it can be conceived as a primitive notion and need not be defined in terms of negation and contradiction.

It is a fact well known to everyone that we must reason, make inferences, and make decisions, in face of contradictions that cannot be simply discharged. Once we accept the idea that contradictory information is to be found in a number of contexts of reasoning, separating inferences that are acceptable from those that are not becomes the task of the logical systems that represent such contexts. The crucial strategy is to weaken the logic in such way that the presence of contradictions does not result in triviality, while at the same time preserving desirable features of the notion of logical consequence. This is indeed a difficult and challenging task. However, if we take a look at real situations in which we have to deal with contradictions, we see that the underlying logic does not have to change very much. Virtually all of the inferences allowed with respect to non-contradictory sentences are still valid. In fact, what actually happens seems to be quite simple: the explosion principle is simply not applied to contradictory sentences. The contradiction is put aside, so to speak, awaiting a later decision.

Logics of formal inconsistency have resources for formally representing situations such as these. They internalize the metatheoretical notion of consistency, expressing it in the object language. Thus we can isolate contradictions in such a way that the validity of explosion is 'restricted' to consistent sentences only, avoiding triviality. This is achieved by means of a modified principle of explosion,

$$\circ A, A, \neg A \vdash B$$

where $\circ A$ means that A is consistent. If we do not have the consistency of A, explosion does not hold with respect to A and $\neg A$:

$$A, \neg A \nvdash B.$$

In the C_n hierarchy, introduced by Newton da Costa (see da Costa, 1963, 1974), the so-called 'well-behavedness' of a formula A, in the sense that it is not the case that A and $\neg A$ hold, is also expressed inside the object language. Thus, in C_1, the first logic of C_n hierarchy, the 'well-behavedness' of A is expressed by A°, as just an abbreviation of $\neg(A \wedge \neg A)$. On the other hand, in $LFIs$, the consistency of a formula A may be a primitive notion, not definable in terms of the non-contradictoriness of A. *LFIs* not only distinguish triviality from contradictoriness, but also makes it possible to distinguish non-contradictoriness from consistency.

There is no commitment to the *truth* of the sentences that express the contradiction. They may well be taken as a provisional situation that can be decided later, at least in principle. In the framework of logics of formal inconsistency, there is a place for contradictions of an epistemic character – we could even say that they are perfectly well-suited to the idea that there are no contradictions in space-time phenomena, nor in mathematical objects, although we sometimes have to reason in the presence of some contradictory information.

In section 2 we present a brief analysis of the nature of logic, aiming at showing that logics of formal inconsistency find their place in the epistemic character of logic as a normative theory of logical consequence. In section 3 we defend a claim about the nature of contradictions that we have to deal with in a number of situations, arguing that they have an epistemic rather than an ontological character. With respect to the existence of real contradictions, although we do not believe in them, we argue that the most reasonable position is to suspend judgment. The discussion in sections 2 and 3 provides support for the discussion in section 4. Three concepts play a special role in paraconsistent logics: contradiction, consistency and negation. In section 4 we try to show that these concepts are polysemic. If one intends to give them a precise treatment, being faithful to the way they are used in informal reasoning, the sciences, and philosophy, then it must be understood that both negation and contradiction may be conceived from either an epistemic or an ontological viewpoint. With regard to consistency, we offer arguments showing that it can be conceived as a primitive notion that is not necessarily definable in terms of negation and contradiction. Finally, in section 5, we compare and establish our position regarding paraconsistency in relation to other approaches.

2   On the nature of logic

2.1   Some remarks on logical pluralism

Logical pluralism is the view that there is more than one legitimate account of logical consequence. To the extent that there are a variety of treatments of the notion of logical consequence, and plenty of logics to suit all tastes, it seems that one could say that logical pluralism is a fact. Accordingly, although some of these logics have a better philosophical justification than others, the question of which one is the correct logic turns out not to be so important. However, a closer look shows that the issue is not that simple. The attempts to justify this plurality of treatments of logical consequence yield some difficulties, the most important one being the threat of relativizing truth. The reason is that, in principle, logical consequence has to do with truth preservation. In any account of logical consequence that is worth studying, we cannot make inferences that can lead from truth to falsity. Consider, for example, two different notions of logical consequence, represented by $\vdash_1$ and $\vdash_2$. It can happen that, from the same set Γ of premises a given sentence A follows and does not follow from Γ, depending on the features of $\vdash_1$ and $\vdash_2$. Now, if the sentences of Γ are true, what is to be said about A? On the one hand it must be true, and on the other it could be false. However, this result seems peculiar since it implies that a given circumstance, expressed by the truth of the premises Γ, does not unequivocally determine the truth of A. This leans dangerously towards assuming that truth is relative, a step we do not want to take.

Provided that the language is the same, if there are different logical consequence relations and all of them are 'correct', then the obvious question is this: how are we to know what logic should be applied in a given situation? It seems that the only available answer is that what follows from what depends on the context: different accounts of logical consequence are designed to deal with different contexts of reasoning that determine which inferences we can or cannot make. This answer, however, is a bit vague and too pragmatic to have philosophical significance. After all, the truth of a sentence should not be a matter of a convenient decision between two different logics, which in turn is a matter of different contexts. Then, a question remains unanswered: what is the feature of logic that allows for different accounts of logical consequence?

2.2 What logic is all about?

In the strict sense, logic is a theory of logical consequence. As such, its task is to formulate principles and methods for establishing when a sentence A follows from a set of premises Γ. The question then is: what are the principles of logic about? Are they about reality, thought, or language? This is a central question if we are to think about the nature of logic. In other words, the question is whether logic has an ontological, an epistemological, or a linguistic character.[1]

---

[1]On the problem of the nature of logic, see, for instance, Chateaubriand (2001, Introduction), Tugendhat & Wolf (1989, chap. 1), and Popper (1963, pp. 206ff).

We will restrict our attention here to the epistemological and the ontological aspects of logic. The linguistic conception of logic, according to which logic is above all a theory about language, has little to add to the issues addressed here.

Throughout the history of philosophy, different answers have been given to this question. Aristotle's defense of the principle of non-contradiction as "the most certain principles of all things" (*Metaphysics* 1005b11) has an ontological character.[2] In this sense, logical principles are about reality, like general laws of nature that apply to any kind of object. This idea is also generally found in Russell and Frege.

On the other hand, in the modern period, logic has had an epistemological character as the study of the rules of correct reasoning. This is in accordance with the epistemological approach that is a central feature of modern philosophy.[3]

However, we would like here to emphasize the epistemic character of intuitionistic logic, which has its motivations in Brouwer's views on mathematics as an activity of the human mind. For Brouwer, logic merely represents the operations of thought in constructing mathematical proofs.[4] According to this position, logic is also descriptive. This, nevertheless, does not make intuitionistic logic vulnerable to the accusations of psychologism advanced by Frege (1893), for at least two reasons. Firstly, it is not the case, as Frege claims, that instead of 'things themselves', logic would be interested only in mere representations, since the point is precisely that the things themselves are, in this case, constructions of thought. Secondly, intuitionistic logic intends to describe *correct* mathematical reasoning, that is, the way the mind proceeds in the construction of *correct* mathematical proofs. Accordingly, it has a normative character in the sense that it says how one should reason in order to reason correctly.

However interesting this discussion about the nature of logic might be, what really matters to us is that it indicates why there may be several accounts of logical consequence. The question of the nature of logic at the present time, in light of logical pluralism, has acquired a character that it did not have during the last century. When, for instance, Frege attacks psychologism, and Quine rejects the ontological character of logic claiming that to ask if logic is "a compendium of the broadest traits of reality" is "unsound; or all sound, signifying nothing" (Quine, 1996, p. 96), the logic

---

[2]All passages from Aristotle referred to here are from Aristotle (1996).

[3]A good example is the *Logic of Port-Royal* (Arnauld & Nicole, 1662, p. 23). It is also clear in Kant, for whom logic was part of a theory of knowledge – see for instance the *Introduction* to the Jäsche Logic (Kant, 1992, pp. 527ff.).

[4]See Brouwer (1907) and the *Disputation* in Heyting (1956).

at stake is the classical one. At present, the question should be no longer whether or not logic has this or that feature, but should become a question of *which* logic has this or that feature. The difference between two accounts of logical consequence may be that one takes mainly an ontological approach, strictly related to truth preservation, while the other deals not only with truth but also with something else – for example, some notion of constructive provability. In order to explain logical pluralism in a coherent way, while at the same time giving an answer of philosophical significance and avoiding making truth relative, it seems to us that a better alternative is to show that different accounts of logical consequence are related to different aspects of logic, especially its epistemic aspects. Let us take a look at the differences between classical, intuitionistic, and paraconsistent logics.

We agree that classical logic may be motivated by a realist conception of truth. Indeed, if truth is independent of mind, and sentences are true or false regardless of the knowledge we have of them, bivalence and the excluded middle are very compelling principles. At the same time, non-contradiction also seems to be very compelling as a principle about reality, along the lines defended by Aristotle. In addition, we see the principle of explosion as an even stronger way of expressing the idea that there is no contradiction in reality, since triviality is unacceptable from any viewpoint. We would like to highlight that the validity of explosion in intuitionistic logic can be plausibly interpreted as saying that there is no contradiction in mathematical objects, even if they are constructions of mind.

Accordingly, classical logic would be the only account of logical consequence that has a predominantly ontological character. It would in fact be difficult to justify that more than one logic operates on the ontological level. The reason is that there is only one world, only one reality. If a theory speaks of reality, there should not be another different theory saying different things about reality. Once we accept that an account of logical consequence has to do primarily with reality, including space-time phenomena and mathematical objects, this account should be unique.

Intuitionistic logic, in its standard formulation (Heyting, 1956), can be considered as a theory of logical consequence of a normative character when we have as a background the notion of constructive proof expressed by the *BHK* interpretation. As Raatikainen (2004) shows, Heyting's and Brouwer's attempts to identify a notion of truth with an intuitionistic notion of provability have not been successful. The reason, in our view, is that intuitionistic logic is not only about truth, but rather it is about truth obtained by constructive means.

With respect to logics of formal inconsistency, we claim that they may be interpreted as normative theories of logical consequence in the sense that they tell us how to reason correctly in the presence of contradictions. Analogously to intuitionistic logic, which can also be understood as a description of how the mind proceeds in the construction of correct mathematical proofs, logics of formal inconsistency also combine a normative and a descriptive aspect. It is a fact that in informal reasoning people deal with contradictions in various situations, but in none of them is the principle of explosion applied, regardless of one's awareness of paraconsistent logics. From this point of view, logics of formal inconsistency are descriptive. They are also normative, of course, since there is no doubt that triviality must be avoided.

Let us return now to the question about the feature of logic that allows different accounts of logical consequence. Logic has an epistemological aspect. It may be, therefore, that it is not only truth that is being accounted for in a theory of what follows from what. We have tried to show that this may be the case in both intuitionistic logic and paraconsistent logics. When we look at the issue of logical pluralism from this point of view, we find a way to accommodate different logics, maintaining truth as a central concept but allowing different approaches to logical consequence.

3 On the nature of contradictions

In the last section we defended the view that logics of formal inconsistency have an epistemic and normative character, and that their aim is to tell us how to draw inferences in the presence of contradictions. Now we will address the question of the nature of contradictions.

Contradictions appear in several situations; for instance, contradictory pieces of information, laws, databases, and scientific theories. One must notice that in the first three cases, contradictions have a clearly epistemic character in the sense that they are provisional and can possibly, at least in principle, be eliminated by means of further investigation. Scientific theories, on the other hand, deserve special attention. An important point about the nature of contradictions lies here, namely, whether there are contradictions that have an ontological character.

3.1 A Kantian inspired view of contradictions in science

It is a fact that scientific research deals with contradictions (see, for instance da Costa & French, 2003, chap. 5). In Physics, it sometimes occurs that two non-contradictory theories yield contradictions when put together. Hence, since contradiction implies triviality in the framework of classical logic, it seems obvious that we need a non-explosive logic in order to give a global account of it. In regard to this matter, it is worth noting that when the physicist considers two incompatible theories, even if it is a matter of revising one of them, (s)he is already reasoning in a paraconsistent way, since (s)he limits the application of the principle of explosion.

Before addressing the question of the nature of contradictions in scientific theories, we will make a brief excursion into the Kantian 'negative results' with respect to human knowledge.

Over than two hundred years ago, Kant presented fundamental insights about the limits of human reason that we essentially endorse. Kant's aim was to give foundations to metaphysics, separating, so to speak, the good metaphysics from the bad. Briefly, he concluded that human reason has limits that should not be surpassed. When such limits are surpassed, error and contradiction may be obtained, which is a sign that something has gone wrong. There is an unbridgeable distinction between reality as it is (things-in-themselves) and reality as it presents itself to us by means of our experience (phenomena). Hence, and this is the point we want to emphasize, there are aspects of reality that are inaccessible to human knowledge.

The path taken by Kant in reaching these conclusions could be questioned. Nevertheless, it is very reasonable to affirm that the problems in Kant's line of reasoning, especially in his account of space and time, are due to the fact that he was working in the framework of the science of his time (one of Kant's goals was to provide foundations for Newtonian physics). In any case, the central point is that Kantian 'negative results' seem to express essential features of human knowledge and its limits.

What we present next, with a frankly Kantian inspiration and based on the account given in da Costa & Krause (2014), is a view of the workings of scientific theories. Scientific research involves three levels: (i) *Reality* (with a capital 'R'), (ii) empirical phenomena, and (iii) scientific theories. *Reality* represents those aspects of reality that are inaccessible to human reason; this idea, of course, corresponds to the Kantian notion of things-in-themselves. By means of empirical phenomena we establish a mediated relationship with *Reality*. It is based on this level that theories are elaborated, since it is here that scientific experiments and data collection take place – everything, of course, mediated by perception, our conceptual apparatus, and measuring instruments. There is not, nor can there be, direct, non-mediated access to *Reality*.

We consider the schema above as the best account of the practice of scientific research and the results of the empirical sciences. We also agree that it is a somewhat skeptical view, but we think it is the right measure of skepticism that a thousand years of philosophical reflection suggest as a plausible and careful position.

Now it may be asked: in the schema above, where do contradictions occur? Of course they occur in theories, since it is a fact that we deal with contradictory theories (cf. da Costa & French (2003, chap. 5), da Costa & Krause (2004), da Costa & Krause (2014)). Besides, it is plausible that contradictions also occur in empirical phenomena.

The central question is whether there are contradictions in *Reality* (with a capital 'R'). These would be what we call *real contradictions*. However, in order to answer this question once and for all we would have to have unconditioned access to *Reality*, and according to the schema presented above this is not possible. The claim that there are contradictions in *Reality* cannot be anything but a conjecture.

Surely, we cannot simply extend the features of empirical phenomena and scientific theories to *Reality*. The occurrence of contradictions in phenomena and theories does not imply the existence of contradictions in *Reality*. Actually, what we have on our hands is a typical philosophical problem that has no prospect of a definite solution. Once we accept that *Reality* is inaccessible, we cannot, from the fact that a given theory has some internal contradictions, decide whether it is *Reality* that is contradictory or if such a theory is going to be corrected in the course of further investigations.

Claiming that there are real contradictions, in the sense explained above, is rather metaphysical, in the bad sense of the term, the kind of speculative philosophy Kant tried to fight against in the eighteenth century. We acknowledge, of course, that we cannot conclusively affirm the opposite, that there are no real contradictions. It would imply that we have access to *Reality*, a refutation of the schema presented above, and we are not so paraconsistent as to endorse simultaneously a view and its refutation.

Notice, however, that in all circumstances in which we have full and unproblematic access to phenomena, there is no sign of contradictions. Assuming that nature is uniform and regular, an assumption that plays an important role in scientific research, it seems that there can be no evidence that there are contradictions in *Reality*. This is the reason why we do not believe in real contradictions. But, again, that is no more than a conjecture that cannot be conclusively established as true.

3.2   A semantics for a logic of formal inconsistency

A question that certainly has already occurred to the reader regards the semantics of logics of formal inconsistency. Indeed, a relevant point is how to provide an interpretation for contradictory sentences.

Like intuitionistic logic, paraconsistent logics were initially introduced exclusively in proof-theoretical terms. Later, bivalued semantics were proposed. As an example, let us consider mbC, a logic of formal inconsistency studied in Carnielli et al. (2007). mbC admits a complete and correct semantics that excludes the simultaneous attribution of the value 0 to a pair of sentences A and $\neg A$, but at the same time allows the simultaneous attribution of the value 1 to a pair of sentences A and $\neg A$. This is achieved by means of the following clause ($\neg$ is paraconsistent negation):

$$v(\neg A) = 0 \quad \text{implies} \quad v(A) = 1.$$

Note that $v(\neg A) = 0$ only specifies a sufficient condition for attributing the value 1 to A, and that $v(A) = 1$ is a necessary condition for $v(\neg A) = 0$. So, although they cannot both receive the value 0, they may receive both the value 1. We have, thus, a situation dual to intuitionistic logic, where there may be a stage t of a Kripke model such that both A and $\neg A$ ($\neg$ here is intuitionistic negation) have the value 0 in t. In Kripke models for intuitionistic logic, 1 and 0 should not be read as *true* and *false*, but rather as *proved* and *neither proved nor refuted*.

Analogously, regarding the above-mentioned bivaluation semantics that allows both A and $\neg A$ to receive 1, the value 1 attributed to a sentence A may be read as 'A is taken to be true', 'A is possibly true', 'A is probably true', or perhaps better as 'there is some evidence that A is true' in the sense of there being reasons for believing that A is true. Thus when we attribute the value 1 to a pair of sentences A and $\neg A$, this does not need to be understood as if we were really claiming that both are true in the strong sense that there is something in the world that makes them true. Rather, A and $\neg A$ are both being taken in a sense weaker than true, perhaps waiting for further investigations that will decide the issue, and discard one of them. Hence, the value 1 that in bivaluation semantics can be attributed to a sentence and to its negation does not mean any commitment to the truth of a contradiction.

Logics of formal inconsistency are not only able to express an epistemic view on the character of contradictions, but also have resources to fully recover classical logic where appropriate. Let us take a closer look at this point.

The so-called *derivability adjustment theorems* (from now on, *DATs*) allow the establishment of a relationship between two logics. In the case of mbC and classical logic, it can be proved that:

For all Γ and for all B, there is a Δ such that:

$\Gamma \vdash_C B$ if and only if $\Gamma, \circ(\Delta) \vdash_{mbC} B$,

where $\circ\Delta = \{\circ A : A \in \Delta\}$ and $\vdash_C$ means classical consequence.

Classical inferences not allowed in mbC, especially those related to paraconsistent negation, may be restored when we add some information to the system. How a DAT works for mbC may be seen from the examples below:

$\nvdash_{mbC} (A \to B) \to ((A \to \neg B) \to \neg A)$

$\nvdash_{mbC} \neg\neg A \to A$

$\nvdash_{mbC} A \to \neg\neg A$

$\nvdash_{mbC} (A \to B) \leftrightarrow (\neg B \to \neg A)$

while

$\circ B \vdash_{mbC} (A \to B) \to ((A \to \neg B) \to \neg A)$

$\circ\neg A \vdash_{mbC} \neg\neg A \to A$

$\circ A \vdash_{mbC} A \to \neg\neg A$

$\circ A, \circ B \vdash_{mbC} (A \to B) \leftrightarrow (\neg B \to \neg A)$

Before closing this section, we would like to make some remarks about a position in paraconsistency that fundamentally differs from our own. According to dialetheism, a philosophical view defended in a number of books and papers by Graham Priest and his collaborators (see, for example, Priest & Berto (2013) and Priest (2006)), there are true contradictions; that is, there are sentences A such that A and $\neg A$ are both true. Saying that there are true contradictions, however, is virtually the same as saying that there are contradictions in the world, or contradictions with ontological character. Indeed, dialetheism maintains that contradictions also occur in the empirical world (cf. Priest, 2006, p. 159).

As we have tried to make clear in the previous sections, we accept contradictions of an epistemic character, and it is not necessary to believe that there are real contradictions in order to devise a paraconsistent logic. However, independently of the personal beliefs of the authors of this paper, the most reasonable position would be to suspend judgment with respect to the existence of real contradictions. With respect to dialetheias, since the contradictions that we do accept have an epistemic character, we consider it inappropriate to speak of true contradictions.

4   On the senses of contradiction, consistency and negation

In this section, we will present some remarks on three concepts central to paraconsistency: contradictions, consistency, and negation. The purpose is to show that these concepts have more than one meaning. This indicates several unexplored possibilities for investigation.

4.1   Contradictions

It is not surprising, when it comes to contradictions, that Hegel is a philosopher who is always recalled. In his works, we find passages that seem to be in perfect opposition with the principle of non-contradiction. In his *Science of Logic*, we read that

> Everything is inherently contradictory, and in the sense that this law in contrast to the others expresses rather the truth and the essential nature of things.
> (Hegel, 1998, p. 439).

He was perhaps the thinker who most emphatically addressed the issue, stating that both thought and the world are contradictory. However, the meaning of 'contradiction' in Hegel is equivocal.

Explaining any concept of Hegel's system in a few words, without appealing to jargon, is not an easy task. It is perhaps appropriate to begin with a warning. We will not become involved here in the details of the idiosyncratic Hegelian terminology, but will simply try to make clear the essential points needed to grasp the ambiguity in the meaning of contradiction in Hegel.

We must first return to the Kantian 'negative results' mentioned in section 3. Kant's analysis establishes an insurmountable gap between subjectivity and objectivity. Scientific knowledge depends on our cognitive apparatus; we know nothing about the objects of experience apart from the conditions by which they are given to us. As an immediate consequence, unconditioned knowledge is not possible. In opposition to Kant, Hegel believes that unconditioned (i.e. absolute) knowledge not only should but *must* be possible. A necessary condition for attaining it is to overcome the gap between subject and object. In Hegelian terminology, nothing can be left 'outside' the absolute, otherwise the absolute would not be really absolute. The point is that the truth must be the whole truth, the absolute truth, and hence must encompass the whole of reality in its temporal development, together with subjectivity, in a unity. But what could this unity be? We can safely say that it includes reality in all of its constituent moments, opposing forces and conflicts, and collapses the distinction between subject and object.[5]

The point is that in Hegel a contradiction also means a mere opposition that is a result of conflicts between different things or 'moments' of the same thing. Hence, strictly speaking, this is not a contradiction in the sense of a violation of the Aristotelian claim that "the same attribute cannot at the same time belong and not belong to the same subject in the same respect" (*Metaphysics* 1005b19-21).

We wish to call attention to the fact that the above interpretation, which identifies a weaker notion of contradiction in Hegel, is closely related to the basic idea of possible-translations semantics (and in particular to society semantics) introduced in Carnielli (1990) and treated in detail in Carnielli et al. (2007).

Indeed, the internal logic of a group of individuals is not necessarily the same as the logic of individuals taken separately. There are different standards of rationality with respect to individuals on the one hand and groups on the other. Non-contradictoriness, for instance, when applied to groups, differs from that applied to individuals, precisely because among the individuals (members of the group) there may be conflicting beliefs, opinions, judgments, etc. Individuals very often act like 'contrary forces' within a society. The whole society, however, is subject to stricter standards of rigor. As a very simple example, suppose a group of referees has to decide upon a selection of papers to be published. Only some of them can be selected, and with respect to each paper the whole group must take a position A or *not A*, but not both. Clearly, in the case of two different referees that defend, respectively, positions A and *not A* for a certain paper,

---

[5]The idea that there are opposing elements existing simultaneously in a whole is found in many places in Hegel's writings. The following passage from Hegel (1977, sec. 20), illustrates this interpretation and the weaker sense of contradiction: "The True is the whole. But the whole is nothing other than the essence consummating itself through its development. Of the Absolute it must be said that it is essentially a *result*, that only in the *end* is it what it truly is; and that precisely in this consists its nature, viz. to be actual, subject, the spontaneous becoming of itself. Though it may seem contradictory that the Absolute should be conceived essentially as a result, it needs little pondering to set this show of contradiction in its true light."

we may well say that there is a contradiction between them, but this is clearly a weaker sense of contradiction. The example above shows that having more than one notion of contradiction can be a useful tool for expressing a number of features in different contexts of reasoning.

4.2 Consistencies

Classical logic works with two different but equivalent notions of consistency, called by Hunter (1973, sec. 24) simple and absolute consistency. Simple consistency is tantamount to non-contradictoriness, and absolute consistency, tantamount to non-triviality. It is clear that in classical logic, due to the validity of the principle of explosion, these two notions are equivalent. Of course, in paraconsistent logics these two notions do not coincide, since the point is precisely to allow some contradictions (i.e. simple inconsistency) without triviality (maintaining absolute consistency). However, as has been noted elsewhere (Carnielli, 2011), in informal reasoning, there are a number of different notions of consistency.

First of all, notice that absolute consistency definitely does not match any informal or intuitive notion of consistency. As pointed out earlier, there are situations where we have to deal with contradictions, and this is perfectly rational and acceptable. Triviality, on the contrary, is unacceptable in any situation. Furthermore, there are intuitive notions of consistency, present in the use of natural language, that do not correspond exactly to the notion of simple consistency. We sometimes use consistency in a sense that is not necessarily related to or defined in terms of negation. In scientific research we sometimes come across information that for some reason is suspicious. This does not *need* to involve a contradiction. Simultaneous but non-conclusive evidence for A and $\neg A$ indicate that both are suspicious. But there may be suspicious sentences not involving any contradiction. The information conveyed by A may be very unlikely, and incoherent with respect to previous data represented by a set Γ of sentences, even though there may be no formal contradiction derivable from A and Γ together. It is not uncommon to say in such cases that A is inconsistent.[6]

Moral dilemmas are another typical situation where a person has to deal with inconsistency, but not necessarily with a formal contradiction. A moral agent may believe that (s)he has moral obligation to perform two actions, A and B, but cannot do both, because they are mutually exclusive.

---

[6]In the logic *mbC*, already mentioned here, the consistency of A, represented by $\circ A$, is not equivalent to non-contradictoriness of A. $\circ A$ implies $\neg(A \wedge \neg A)$, but the converse does not hold.

We may well say that the belief of the agent is inconsistent, in a very reasonable sense of inconsistency. Let us take as an example the well-known dilemma, posed by Sartre (2007), of the man in occupied France who on the one hand wants to fight the Nazis but on the other must take care of his mother. He believes that each alternative is a moral obligation, and that doing one implies not doing the other. However, he may well believe that he has both obligations, while not believing a formal contradiction in the strict sense, i.e. not believing that he has and does not have the obligation of doing something. In order to obtain a formal contradiction, we need some other principles of deontic logic that the agent may be not aware of.

These notions of consistency should be independent of model-theoretical and proof-theoretical means. According to Field (1991), it would be possible to treat a notion of logical consistency as a primitive metalogical notion not reducible to semantic consistency or syntactic consistency. However, it seems that his efforts were not successful, as Wojtowicz (2001) points out; Field's attempts essentially amount to abbreviations for statements about the consistency of some metatheoretical sentences. This shows that treating consistency as independent from model-theoretical and proof-theoretical means is not so obvious. It seems that a better approach is to treat the question *more geometrico*, axiomatizing the notion of consistency as a primitive one, but making it relate to negation and other logical connectives in an appropriate way, as it is done in *mbC* and other *LFIs* (cf. Bueno-Soler & Carnielli, 2014).

4.3   Negations

A point that sometimes causes uneasiness at first sight is that logics of formal inconsistency have more than one negation. This is strange, one might say, because it seems that a negation without the features of classical negation would not properly be a negation. Our reply is that this uneasiness is mistaken. It is quite possible that this uneasy feeling may be caused by a realist conception of truth, connected to the meaning of classical negation, even if one is not completely aware of it. However, we can safely affirm that there are at least two senses of negation, one ontological and one epistemic, corresponding to the two aspects of logic we tried to distinguish in section 2.

We do use, in informal reasoning, a negation that is weaker than the classical one and that arguably has an epistemic character. It happens, for instance, when we say 'not A', but are not sure about how much we are denying – or, in other words, when we are not completely sure that A is consistent (or solid, well-established, etc.). In this case, we write $\neg A$ where

¬ means paraconsistent negation. On the other hand, suppose that 'A is not the case' has already been confirmed, and we can safely say that it has been well established. Now, we affirm 'not A', but this time in a stronger sense, employing a stronger negation. We are now making an assertion that intends to say something in the framework of classical logic, and we write ∼A where ∼ means classical negation. What occurs, in this case, is that in conclusively establishing ∼A we have also established the consistency of A. When we write ∼A we also mean that we cannot have A (a contradiction of the form $A \land \sim A$ is unacceptable, like in classical logic). On the other hand, in writing ¬A we still leave open the possibility that the conclusion may be revised, i.e. that we are not completely convinced that A is not the case. Notice that it is not by chance that epistemic notions (such as to be convinced, to establish conclusively, etc.) occur here. Thus, it seems very reasonable to consider that classical and paraconsistent negation have, respectively, an ontological and an epistemic character. Of course, the distinction we have made in section 3, namely, that a contradiction may have an epistemic or an ontological character, also naturally applies to the negation used to express the contradiction. Even if one admits that negation is unique from an ontological viewpoint (a reasonable position, since there should not be more than one ontology), from the epistemic viewpoint we have at least one additional negation, precisely the negation that occurs in contradictions that have an epistemic character. Furthermore, nothing prevents us from having still more negations.

Now we may take consistency as a primitive operator, ∘, which is justifiable once we acknowledge that consistency does not coincide with triviality, that there is more than one sense of consistency, and that consistency is not always intuitively explained in terms of negation. We may also take as primitive a paraconsistent negation, which, from the epistemological point of view, antecedes classical negation. With all of this, in *mbC* we may define a stronger negation in the following way:

$$\sim A := A \to (\circ A \land A \land \neg A).$$

It may be proved in *mbC* that ∼ so defined has all the properties of classical negation.[7]

---

[7] With the help of this definition, $A \lor \sim A$ and $A \to (\sim A \to B)$ may be proved in *mbC*. Since *mbC* is an extension of intuitionistic sentential positive logic (see Carnielli et al., 2007), classical logic may be restored within *mbC*. Notice the difference between, on the one hand, restoring classical consequence by means of a definition of a classical negation and, on the other, doing so by means of a *DAT* (cf. section 2). In the latter case, the point is the information that has to be available in order to restore classical logic. The former shows that, in a certain sense, although the idea is to restrict inferences valid classically, *mbC* is an extension of classical logic.

To sum up, suppose that we have some grounds, working with some data and previous results, for considering that A is not the case. But we still have doubts with respect to the previous results; in other words, we have not yet found that A is consistent, and so write $\neg A$. Once we establish, by whatever means, that A is indeed not the case, we can now use the classical, stronger negation to affirm not A, i.e. $\sim A$.

The fact that negation is an ambiguous notion that sometimes has a weaker sense is not fully understood, and occasionally causes mistaken reactions against paraconsistency. Slater (1995) argues against Priest's paraconsistent system LP (Priest, 1979), but his criticism supposedly holds for paraconsistent logics in general. He appeals to the traditional notions of sentences being subcontraries but not contradictories, claiming that paraconsistent negations are not 'real' negations but a kind of subcontrariety operator. We say that two sentences A and B are subcontraries when they cannot be both false but can be both true. Of course, if paraconsistent negation were only a way to express subcontrariness, it would hardly be of any philosophical significance.

In Béziau (2002) and Béziau (2006), we find an effective logical and philosophical defense of paraconsistent logics against Slater's arguments, but we think that there are still some remarks to be made on this issue from the point of view of logics of formal inconsistency.

We have just seen above how we can go from a paraconsistent negation to a strong negation that has all the properties of classical negation. It seems to us that it makes no sense to suppose that paraconsistent negation is a kind of subcontrariety operator, for otherwise classical negation would not be definable from paraconsistent negation. More precisely, it is hard to see how paraconsistent negation, as it is conceived in mbC for example, could be a kind of subcontrariety sentential operator, in principle simply allowing the attribution of the value 1 to A and $\neg A$, but at the same time being able to define classical negation based on the conjunction of A, $\neg A$, and $\circ A$. We saw that in mbC consistency is primitive, not related to negation, although we need the help of the consistency operator to define classical negation. How could the property of subcontrariety between A and $\neg A$ be turned into a classical contradictoriness between A and $\sim A$ if the role of $\neg$ were just to form subcontraries?

Even if we accept that $\neg$ is a subcontrariety operator in a sense similar to the sense one could say that intuitionistic negation is a kind of contrariety operator, the only conclusion would be that a paraconsistent negation that allows the attribution of 1 to A and $\neg A$ is a subcontrariety operator in a secondary sense. It seems to us that Slater understands subcontrariety and contradictoriness as strictly related to a notion of truth that complies with the properties of classical negation. But the non-classical negation of *mbC*, as we showed here, is better understood as expressing a notion weaker than truth.

5 Types of paraconsistency

Paraconsistency may be defined as the study of contradictory but non-trivial theories, both in its technical and philosophical aspects. A necessary condition for a logic to be paraconsistent, as we have seen, is the invalidity of the principle of explosion, but explosion may be rejected for several different reasons, related to different attitudes with respect to the nature of contradictions. In fact, there are several types of paraconsistency on the market.

Beal and Restall (2006 pp. 79ff), paraphrasing Quine, list 'four grades of paraconsistency involvement'. The first grade, that is the weaker, is simply dissatisfaction with explosion as a valid inference. This is the case of the relevantists, whose motivation is to avoid the so-called paradoxes of material implication, one of which is precisely the explosion principle. Indeed, relevant logics are a kind of paraconsistent logic, but there is no metaphysical discussion about the nature of contradictions. The point is that there may be no connection between the meanings of the premises and the conclusion. So, if one concludes that 'Descartes is French' from 'Aristotle is Greek and Aristotle is not Greek', since the premises have no connection with the conclusion, i.e. they are not relevant, the inference is thus rejected.

The second grade is the view according to which there are some interesting and contradictory but non-trivial theories. They give as examples of such theories naive set theory and naive truth theory. It is well known that these theories yield contradictions as theorems because of Russell's set and the Liar sentence. There are two remarks to be made here. First, it seems that this position just leaves unnoticed the question of the nature of contradictions. Second, they do not mention the important fact that there are interesting contradictory but non-trivial empirical theories.

The third grade is a somewhat stronger position: some of the given inconsistent but non-trivial theories may be true, that is, it is *possible* that real contradictions exist. This position has already been defended by Newton da Costa.[8]

---

[8]Cf. da Costa (2008, pp. 147 and 237): "[A]t the macroscopic level, the experience seems to indicate that there are no contradictions; however, at a microscopic level, there is nothing to prevent real contradictions. (...) [R]eal contradictions are not impossible, although there is nothing so far proving that they exist."

The forth grade is dialetheism: the thesis that there exist inconsistent but non-trivial theories that are true. These theories truly describe reality by means of some contradictory sentences. This is the position defended by Graham Priest in several places.

To the extent that we suspend judgement with respect to real contradictions, it may seem that our position is similar to the third grade. But that would be a mistake. We do not agree that real contradictions are possible, rather, we only admit that we cannot prove that they are not possible. The alternatives listed by Restall and Beal are not exhaustive, and our position does not fit any of them. We think that a very important position is the one that accepts the existence of epistemic contradictions, the view defended by us here. This view is not new, and has already appeared in the history of philosophy.

Maybe one of the best places to understand the distinction between real and epistemic contradictions is in the response given by Hegel to Kant, who took for granted that contradictions are a signal of error and, as such, can only be produced by reason. Hegel disagrees:

> [According to Kant] The stain of contradiction ought not to be in the essence of what is in the world; it has to belong *only* to thinking reason, to the *essence* of the *spirit* [mind]. It is not considered at all objectionable that the world *as it appears* shows contradictions to the spirit [mind] that observes it; the way the world is for subjective spirit, for *sensibility*, and for the *understanding*, is the world as it appears. (...) [But the true significance of the Kantian antinomies is that] Everything actual contains opposed determinations within it, and in consequence the cognition and, more exactly, the comprehension of an object amounts precisely to our becoming conscious of it as a concrete unity of opposed determinations. (Hegel, 1991, pp. 91 ff)

It is not one-hundred per cent clear that what Hegel means by 'determination' could be represented by a unary predicate of a first order language. Hence, it is also not one hundred per cent clear that opposed determinations would be a violation of the principle of non-contradiction. Furthermore, the reading according to which contradictions in Hegel are due to the ongoing motion of reality is very plausible, but this is not necessarily a violation of the principle of non-contradiction.[9] What we want

---

[9] Aristotle was quite clear in saying that an object having different properties at different moments of time, or seen from different perspectives, would not be a counterexample for the principle of non-contradiction (cf. *Metaphysics*, 1009b1 and 1010b10).

to call attention is how clearly Hegel explains the view, which he rejects, that contradictions belong, or are yielded by, 'thinking reason'.

Hegel's description of Kant's position is very close to what we understand by epistemic contradictions: those produced by limitations in our cognitive apparatus, flaws in the instruments used in experiments, the inability of the available theories to deal with data at hand – in a word, contradictions do not originate in the world itself, but in *the way the world appears to the mind that observes it*. Accordingly, the rejection of the principle of explosion and the acceptance of *some* epistemic contradictions by paraconsistent logics may be understood as a tool for dealing with pairs of contradictory sentences that should not be taken as true but, on the other hand, cannot be simply thrown away. This is our position with respect to paraconsistency, a position that cannot be accommodated in any of the 'grades of paraconsistency' mentioned by Restall and Beal.

5.1 A brief digression on the notion of evidence in paraconsistency

A pair of contradictory sentences may be understood as conflicting information (or evidence, verisimilitude, possibility etc.) – a number of concepts that we deal with in informal reasoning, which makes it very natural to devise a logic designed to give an account of them. Among the possible intuitive interpretations of the acceptance of a contradictory pair of sentences without implying triviality, the notion of *evidence* is particularly promising for paraconsistent logics.

Evidence may be understood, as it is usual in epistemology, as what is relevant for justified belief (cf. Kelly, 2014). Justified belief, in its turn, is traditionally a necessary condition for knowledge. From this point of view, the idea of preservation of evidence presents itself as a topic to be further developed in logics of formal inconsistency. The idea is that, as much as the *BHK* interpretation for intuitionistic logic deals with preservation of (some sense of) construction, a set of inference rules and/or axioms that preserve (some sense of) evidence can be also established.

However, on the other hand, a paraconsistentist approach to evidence may also be done from a probabilistic point of view. The aim of evidence theory is to give an account of reasoning under *epistemic* uncertainty, which may occur due to a lack of data and/or a defective understanding of the available data related to some scenarios. It is distinguished from *randomic* uncertainty, which is uncertainty due to the inherently aleatory character of some scenarios. One of the key points of evidence theory is that in imprecise events, uncertainty about an event can be quantified by the maximum and minimum probabilities of that event. This directly connects evidence theory to probability theory, and therefore to Bayesian conditionalization.

In Bueno-Soler & Carnielli (forthcoming), the authors investigate a paraconsistent approach to theory of probability based on the logics of formal inconsistency. The paper shows that *LFIs* naturally encode an extension of the notion of probability able to express the notion of probabilistic reasoning under contradictions by means of appropriate notions of conditional probability and paraconsistent updating, by means of a version of Bayes' Theorem for conditionalization. The authors argue that the dissimilarity between the notions of inconsistency and contradiction plays a central role in an extended notion of probability that supports contradictory reasoning.

Actually, evidence theory and Bayesian theory of subjective probability are simultaneously generalized by Dempster-Shafer theory, which refers to belief and plausibility. Beliefs from different sources can be combined (by means of the so-called Dempster rule of combination) with various operators to model specific situations of belief diffusion. An analogous treatment can be effected by starting from paraconsistent probability.

Another approach for an evidence interpretation to paraconsistency, related (but not coincident) to paraconsistent probability, is the game-theoretiical view on paraconsistency, still to be developed in full. Carnielli & Rahman (2000) have presented a game-theoretical account (by means of dialogue logic) of paraconsistency. Dialogue logic makes it possible to accommodate the occurrence of contradictions in two (or more) persons' reasoning, and positively contribute to the debates concerning the ontological versus epistemological nature of contradictions. Rahman and Carnielli's paper has attracted considerable attention (cf. e.g. Bendegem, 2011) as an important first step in reformulating paraconsistent logic in in terms of dialogue format.

Dialogue logic is not the only approach to paraconsistency from the point of view of game theory. For instance, in a paraconsistent game-theoretical scenario the truth of a sentence can be defined in terms of the lack of winning strategies for the opponent, instead of in terms of existence of winning strategies of the proponent; a similar view is defended in Galliani (2014), but with different assumptions (namely, the existence of true contradictions or dialetheas). A related possibility is granted by the definition of the so-called 'team semantics' in Väänänen (2007), although the idea of a society producing semantics had been introduced (years before) in Carnielli & Lima-Marques (1999).

6 Final remarks

A logic has epistemic rather than ontological character when its subject matter is not only truth, but some concept strictly related to reason. This is the case for intuitionistic logic, which is concerned with truth attained in a specific way, by means of a constructive proof. We also claim that this is a way for understanding paraconsistency in general, and particularly logics of formal inconsistency. The latter, we may say, is concerned with truth, since classical logic can be recovered with respect to consistent sentences, but it is also concerned with a notion weaker than truth, and it is precisely this weaker-than-truth notion that allows an intuitive and plausible understanding of the acceptance of contradictions in some contexts of reasoning.

Intuitionistic logic is a special case of paracomplete logics, that is, logics in which there is a model M and a sentence A such that both A and its negation do not hold in M. Mathematicians deal with lack of information in the sense that there are many unsolved mathematical problems. This is one of the reasons for the rejection of excluded middle by intuitionists. In empirical sciences, on the other hand, although obviously many things are not known, the researcher sometimes deals with conflicting information, and very often with contradictory information. Thus (s)he sometimes may have to provisionally consider two contradictory claims, one of which in the due time will be rejected. Both intuitionistic and paraconsistent logics may be conceived as normative theories of logical consequence with epistemic character. Moreover, both are also descriptive: the first, according to Brouwer and Heyting, intends to represent how the mind works in constructing correct mathematical proofs, while the latter, we argue, represents how we draw inferences correctly when faced with contradictions.

In fact, it is not surprising that we find a kind of duality in the motivations for intuitionistic and paraconsistent logics. Brunner & Carnielli (2005) showed that dual-intuitionistic logics are paraconsistent, and hinted at similar results in the other direction. Nevertheless, the dual of Heyting's well-known intuitionistic logic gives rise to a new paraconsistent logic, that is, one that is not a familiar paraconsistent logic. These results, in any case, show an intrinsic relationship between both paradigms, but there is of course much to be investigated in this regard.

In section 4 we argued that negation, contradiction and consistency, concepts central to paraconsistency, are polysemic. Again, this opens many possibilities for investigation. Conceptions of consistency as coherence of a group of statements, or as continuity in time of holding a group of statements, could also be studied from a paraconsistent point of view. It is worth noting that the ideas presented here indicate that there is much more to be explored in Hegel than just the (perhaps misleading) idea of true contradictions.

Although classical first-order logic is a very powerful tool for modelling reasoning, the fact that it does not handle contradictions in a sensible way, due to the explosion principle, is an essential drawback. In theoretical computer science, for example, facts and rules of knowledge bases, as well as integrity constraints, can produce contradictions when combined, even if they are sound when separate, and this is also the case for description logics. For this reason, the development of paraconsistent tools has turned out to be an important issue in working with description logics and expandable knowledge bases, as well as with large or combined databases (see Carnielli et al. (2000) and Grant & Subrahmanian (2000)).

The new approach to consistency granted by the logics of formal inconsistency has also generated a good deal of interest in the field of inferential probability and confirmation theory. Due to the well-known debate in the philosophical literature on the long-standing confusion about probability when confronted with confirmation (cf. e.g. Fitelson (2012) and Fitelson (2006)), notions such as coherence, credence, individual consistent or coherent profiles versus group profiles, etc. (see Fitelson, 2007), can be fruitfully approached from the point of view of the logics of formal inconsistency, although there is still a long way to go.

We conclude with a few words on a famous remark made by the Persian philosopher and polymath Ibn Sina, also known as Avicenna. For those who insist in denying the principle of non-contradiction, he responded virulently:

> As for the obdurate, he must be subjected to the conflagration of fire, since "fire" and "not fire" are one. Pain must be inflicted on him through beating, since "pain" and "no pain" are one. And he must be denied food and drink, since eating and drinking and the abstention from both are one [and the same]. (Avicenna, 2005, book I, ch. 8, p. 43)

It is not to be believed that Avicenna, a man versed in astronomy, alchemy, psychology, Islamic theology, logic, mathematics, physics, poetry, and medicine would like to see opponents of the law of non-contradiction being beaten and burned. But if so, perhaps nowadays Avicenna would recognize that paraconsistentists (at least most of them) are not saying that to be burned is the same as not to be burned. Rather, they are saying that to be and not to be (burned, beaten, or anything else), as we have tried to show here, sometimes have to be considered together as a sort of rational possibility.

Acknowledgements

The first author acknowledges support of *FAPESP* (São Paulo Research Council) and *CNPq*, Brazil (The National Council for Scientific and Technological Development). The second author acknowledges support from the Universidade Federal de Minas Gerais (edital 08/2010) and *FAPEMIG* (Fundação de Amparo à Pesquisa do Estado de Minas Gerais, research project 21308). Both authors wish to thank Tadeu Verza for the reference on Avicenna.

References

Aristotle (1996), *The Complete Works of Aristotle*, Oxford University Press.

Arnauld, A. & Nicole, P. (1662), *Logic or the Art of Thinking*, Cambridge University Press (1996).

Avicenna (2005), *The Metaphysics of the Healing (transl. Michael E. Marmura)*, Brigham Young University.

Bendegem, J. (2011), 'Paraconsistency and dialogue logic: critical examination and further explorations', *Synthese* **127**, 35–55.

Béziau, J.-Y. (2002), Are paraconsistent negations negations?, *in* W. et al. (eds), ed., 'Paraconsistency: the logical way to the inconsistent', Marcel Dekker, New York, pp. 465–486.

Béziau, J.-Y. (2006), 'Paraconsistent logic! (a reply to Slater)', *Sorites* pp. 17–25.

Brouwer, L. (1907), On the foundations of mathematics, *in* 'Collected Works vol. I. (ed. A. Heyting)', North-Holland Publishing Company (1975).

Brunner, A. & Carnielli, W. (2005), 'Anti-intuitionism and paraconsistency', *Journal of Applied Logic* **3**, 161–184.

Bueno-Soler, J. & Carnielli, W. (2014), 'Experimenting with consistency', *CLE e-Prints*.

Bueno-Soler, J. & Carnielli, W. (forthcoming), 'May be and may be not: paraconsistent probabilities from the LFI viewpoint.'.

Carnielli, W. (1990), Many-valued logics and plausible reasoning, in 'Proceedings of Conference on Many-Valued Logics,Charlotte, USA'.

Carnielli, W. (2011), 'The single-minded pursuit of consistency and its weakness', *Studia Logica* **97**, 81–100.

Carnielli, W. & Lima-Marques, M. (1999), Society semantics and multiple-valued logics, in 'Advances in Contemporary Logic and Computer Science', American Mathematical Society, Providence, pp. 33–52.

Carnielli, W. & Marcos, J. (2001), A taxonomy of C-systems, in W. Carnielli, M. Coniglio & I. D'Ottaviano, eds, 'Paraconsistency: The Logical Way to the Inconsistent, Proceedings of the Second World Congress on Paraconsistency', Marcel Dekker, New York.

Carnielli, W. & Rahman, S. (2000), 'The dialogical approach to paraconsistency', *Synthese* **125**, 201–232.

Carnielli, W., Coniglio, M. & Marcos, J. (2007), Logics of formal inconsistency, in G. . Guenthner, ed., 'Handbook of Philosophical Logic', Vol. 14, Springer.

Carnielli, W., Marcos, J. & Amo, S. D. (2000), 'Formal inconsistency and evolutionary databases', *Logic and Logical Philosophy* **8**, 115–152.

Chateaubriand, O. (2001), *Logical Forms vol. 1*, Campinas: UNICAMP-CLE.

da Costa, N. (1963), *Sistemas Formais Inconsistentes*, Curitiba: Editora da UFPR (1993).

da Costa, N. (1974), 'On the theory of inconsistent formal systems', *Notre Dame Journal of Formal Logic* **XV, number 4**, 497–510.

da Costa, N. (2008), *Ensaio sobre os fundamentos da lógica*, São Paulo, Hucitec.

da Costa, N. & French, S. (2003), *Science and Partial Truth: A Unitary Approach to Models and Scientific Reasoning*, Oxford: Oxford University Press.

da Costa, N. & Krause, D. (2004), Complementarity and paraconsistency, *in* 'Logic, Epistemology, and the Unity of Science', Vol. 1, Dordrecht: Kluwer Academic Publishers, pp. 557–568. S. Rahman; J. Symons; D.M. Gabbay; J.P. van Bendegen. (Org.).

da Costa, N. & Krause, D. (2014), 'Physics, inconsistency, and quasi-truth', *Synthese* **191**, 3041–3055.

Field, H. (1991), 'Metalogic and modality', *Philosophical Studies* **62**, 1–22.

Fitelson, B. (2006), 'Logical foundations of evidential support', *Philosophy of Science* **73**, 500–512.

Fitelson, B. (2007), 'Likelihoodism, bayesiansim, and relational confimation', *Synthese* **156**, 473–489.

Fitelson, B. (2012), 'Accuracy, language dependence, and joyce's argument for probabilism', *Philosophy of Science* **79**, 167–174.

Frege, G. (1893), *The Basic Laws of Arithmetic*, University of California Press (1964).

Galliani, P. (2014), Dialetheism, game theoretic semantics, and paraconsistent team semantics, Pre-print version, Arxiv, 2014 http://arxiv.org/abs/1206.6573.

Grant, J. & Subrahmanian, V. S. (2000), 'Applications of paraconsistency in data and knowledge bases', *Synthese* **125**, 121–132.

Hegel, G. (1977), *Phenomenology of Spirit*, Oxford University Press. (transl. A.V. Miller).

Hegel, G. (1991), *The Encyclopaedia Logic*, Indianapolis, Hacket. (trans. T. F. Geraets, W. A. Suchting, and H. S. Harris).

Hegel, G. (1998), *Science of Logic*, Humanity Books, New York. (transl. by A.V. Miller).

Heyting, A. (1956), *Intuitionism: an Introduction*, London: North-Holland Publishing Company.

Hunter, G. (1973), *Metalogic*, University of California Press.

Kant, I. (1992), *Lectures on Logic*, Cambridge University Press (ed. Young, J.M).

Kelly, T. (2014), 'Evidence', *Stanford Encyclopedia of Philosophy*.

Popper, K. (1963), *Conjectures and Refutations*, New York, Harper.

Priest, G. (1979), 'Logic of paradox', *Journal of Philosophical Logic* **8**, 219–241.

Priest, G. (2006), *In Contradiction*, Oxford University Press (2o. ed.).

Priest, G. & Berto, F. (2013), 'Dialetheism', *Stanford Encyclopedia of Philosophy*.

Quine, W. (1996), *Philosophy of Logic*, 2nd ed., Harvard University Press.

Raatikainen, P. (2004), 'Conceptions of truth in intuitionism', *History and Philosophy of Logic* **25**, 131–145.

Sartre, J. (2007), *Existentialism is a humanism*, Yale University Press. (transl. Arlette Elkaim-Sartre).

Slater, B. (1995), 'Paraconsistent logics?', *Journal of Philosophical Logic* **24**, 451–454.

Tugendhat, E. & Wolf, U. (1989), *Logisch-semantische Propädeutik*, Reclam, Ditzingen.

Väänänen, J. (2007), *Dependence Logic*, Cambridge University Press.

Wojtowicz, K. (2001), 'Some remarks on Hartry Field's notion of "logical consistency"', *Logic and Logical Philosophy* **9**, 199–212.

On the definition of the classical connectives and quantifiers[‡]

Gilles Dowek

* INRIA, 23 avenue d'Italie, CS 81321, 75214 Paris Cedex 13, France
gilles.dowek@inria.fr

Abstract

Classical logic is embedded into constructive logic, through a definition of the classical connectives and quantifiers in terms of the constructive ones.

The history of the notion of constructivity started with a dispute on the deduction rules that one should or should not use to prove a theorem. Depending on the rules accepted by the ones and the others, the proposition $P \vee \neg P$, for instance, had a proof or not.

A less controversial situation was reached with a classification of proofs, and it became possible to agree that this proposition had a classical proof but no constructive proof.

An alternative is to use the idea of Hilbert and Poincaré that axioms and deduction rules define the meaning of the symbols of the language and it is then possible to explain that some judge the proposition $P \vee \neg P$ true and others do not because they do not assign the same meaning to the symbols $\vee$, $\neg$, etc. The need to distinguish several meanings of a common word is usual in mathematics. For instance the proposition "there exists a number x such that $2x = 1$" is true or false depending on whether the word "number" means "natural number" or "real number". Even for logical connectives, the word "or" has to be disambiguated into inclusive and exclusive.

Taking this idea seriously, we should not say that the proposition $P \vee \neg P$ has a classical proof but no constructive proof, but we should say that the proposition $P \vee^c \neg^c P$ has a proof and the proposition $P \vee \neg P$ does not, that is we should introduce two symbols for each connective and quantifier, for instance a symbol $\vee$ for the constructive disjunction and a symbol $\vee^c$ for

[‡]The author wishes to thank Olivier Hermant and Sara Negri for many helpful comments on a previous version of this paper.

the classical one, instead of introducing two judgments: "has a classical proof" and "has a constructive proof". We should also be able to address the question of the provability of mixed propositions and, for instance, express that the proposition $(\neg(P \wedge Q)) \Rightarrow (\neg P \vee^c \neg Q)$ has a proof.

The idea that the meaning of connectives and quantifiers is expressed by the deduction rules leads to propose a logic containing all the constructive and classical connectives and quantifiers and deduction rules such that a proposition containing only constructive connectives and quantifiers has a proof in this logic if and only if it has a proof in constructive logic and a proposition containing only classical connectives and quantifiers has a proof in this logic if and only if it has a proof in classical logic. Such a logic containing classical, constructive, and also linear, connectives and quantifiers has been proposed by J.-Y. Girard (Girard, 1993). This logic is a sequent calculus with *unified sequents* that contain a linear zone and a classical zone and rules treating differently propositions depending on the zone they belong.

Our goal in this paper is slightly different, as we want to define the meaning of a small set of primitive connectives and quantifiers with deduction rules and define the others explicitly, in the same way the exclusive or is explicitly defined in terms of conjunction, disjunction and negation: $A \oplus B = (A \wedge \neg B) \vee (\neg A \wedge B)$. A first step in this direction has been made by Gödel (Gödel, 1932) who defined a translation of constructive logic into classical logic, and Kolmogorov (Kolmogorov, 1925), Gödel (Gödel, 1933), and Gentzen (Gentzen, 1969) who defined a translation of classical logic into constructive logic. As the first translation requires a modal operator, we shall focus on the second. This leads to consider constructive connectives and quantifiers as primitive and search for definitions of the classical ones. Thus, we want to define classical connectives and quantifiers $\top^c$, $\perp^c$, $\neg^c$, $\wedge^c$, $\vee^c$, $\Rightarrow^c$, $\forall^c$, and $\exists^c$ and embed classical propositions into constructive logic with a function $\| \ \|$ defined as follows.

Definition 1

- $\|P\| = P$ *if* P *is an atomic proposition*

- $\|\top\| = \top^c$

- $\|\perp\| = \perp^c$

- $\|\neg A\| = \neg^c \|A\|$

- $\|A \wedge B\| = \|A\| \wedge^c \|B\|$

- $\|A \vee B\| = \|A\| \vee^c \|B\|$

- $\|A \Rightarrow B\| = \|A\| \Rightarrow^c \|B\|$

- $\|\forall x \; A\| = \forall^c x \; \|A\|$

- $\|\exists x \; A\| = \exists^c x \; \|A\|$

If $\Gamma = A_1, ..., A_n$ is a multiset of propositions, we write $\|\Gamma\|$ for the multiset $\|A_1\|, ..., \|A_n\|$.

Kolmogorov-Gödel-Gentzen translation can be defined as follows

- $(P)' = \neg\neg P$, if P is an atomic proposition

- $(\top)' = \neg\neg\top$

- $(\bot)' = \neg\neg\bot$

- $(\neg A)' = \neg\neg\neg(A)'$

- $(A \wedge B)' = \neg\neg((A)' \wedge (B)')$

- $(A \vee B)' = \neg\neg((A)' \vee (B)')$

- $(A \Rightarrow B)' = \neg\neg((A)' \Rightarrow (B)')$

- $(\forall x \; A)' = \neg\neg(\forall x \; (A)')$

- $(\exists x \; A)' = \neg\neg(\exists x \; (A)')$

or more succinctly as

- $(P)' = \neg\neg P$, if P is an atomic proposition

- $(*)' = \neg\neg*$, if $*$ is a zero-ary connective

- $(*A)' = \neg\neg(*(A)')$, if $*$ is a unary connective

- $(A * B)' = \neg\neg((A)' * (B)')$, if $*$ is a binary connective

- $(*x \; A)' = \neg\neg(*x \; (A)')$, if $*$ is a quantifier

For instance
$$(P \vee \neg P)' = \neg\neg(\neg\neg P \vee \neg\neg\neg\neg\neg\neg P)$$

And it is routine to prove that a proposition A has a classical proof if and only if the proposition $(A)'$ has a constructive one.

But, this translation does not exactly provide a definition of the classical connectives and quantifiers in terms of the constructive ones, because an atomic proposition P is translated as $\neg\neg P$, while in a translation induced by a definition of the classical connective and quantifiers, an atomic proposition P must be translated as P.

Thus, to view Kolmogorov-Gödel-Gentzen translation as a definition, we would need also to introduce a proposition symbol P^c defined by $P^c = \neg\neg P$. But this would lead us too far: we want to introduce constructive and classical versions of the logical symbols—the connectives and the quantifiers—but not of the non logical ones, such as the predicate symbols.

If we take the definition

- $\neg^c A = \neg\neg\neg A$

- $A \vee^c B = \neg\neg(A \vee B)$

- etc.

where a double negation is put before each connective and quantifier, then the proposition $P \vee^c \neg^c P$ is $\neg\neg(P \vee \neg\neg\neg P)$ where, compared to the Kolmogorov-Gödel-Gentzen translation, the double negations in front of atomic propositions are missing. Another translation introduced by L. Allali and O. Hermant (Allali & Hermant, 2010) leads to the definition

- $\neg^c A = \neg\neg\neg A$

- $A \vee^c B = (\neg\neg A) \vee (\neg\neg B)$

- etc.

where double negations are put after, and not before, each connective and quantifier. The proposition $P \vee^c \neg^c P$ is then $\neg\neg P \vee \neg\neg\neg\neg\neg P$, where the double negation at the top of the proposition is missing. Using this translation Allali and Hermant prove that the proposition A has a classical proof if and only if the proposition $\neg\neg \|A\|$ has a constructive one and they introduce another provability judgment expressing that the proposition $\neg\neg A$ has a constructive proof. This also would lead us too far: in our logic, we want a single judgment "A has a proof" expressing that A has a constructive proof, and not to introduce a second judgment, whether it is "A has a classical proof" or "$\neg\neg A$ has a proof".

In order to do so, we define the classical connectives and quantifiers by introducing double negations both before and after each symbol.

Definition 2 (Classical connectives and quantifiers)

- $\top^c = \neg\neg\top$

- $\bot^c = \neg\neg\bot$

- $\neg^c A = \neg\neg\neg\neg\neg\neg A$

- $A \wedge^c B = \neg\neg((\neg\neg A) \wedge (\neg\neg B))$

- $A \vee^c B = \neg\neg((\neg\neg A) \vee (\neg\neg B))$

- $A \Rightarrow^c B = \neg\neg((\neg\neg A) \Rightarrow (\neg\neg B))$

- $\forall^c x\, A = \neg\neg(\forall x\, (\neg\neg A))$

- $\exists^c x\, A = \neg\neg(\exists x\, (\neg\neg A))$

Notice that the propositions $\top \Leftrightarrow \top^c$, $\bot \Leftrightarrow \bot^c$, and $\neg A \Leftrightarrow \neg^c A$ where $A \Leftrightarrow B$ is defined as $(A \Rightarrow B) \wedge (B \Rightarrow A)$, have proofs. Thus, the symbols $\top^c$, $\bot^c$, and $\neg^c$ could be just defined as $\top$, $\bot$, and $\neg$.

With this definition, neither the double negations in front of atomic propositions nor those at the top of the proposition are missing. The price to pay is to have four negations instead of two in many places, but this is not harmful.

Yet, there is still a problem with the translation of atomic propositions: as with any definition based translation, the atomic proposition P alone is translated as P and not as $\neg\neg P$. Thus, the property that a sequent $\Gamma \vdash A$ has a classical proof if and only if the sequent $\|\Gamma\| \vdash \|A\|$ has a constructive one only holds when A is not atomic. For instance, the sequent $P \wedge^c Q \vdash P$, that is $\neg\neg((\neg\neg P) \wedge (\neg\neg Q)) \vdash P$, does not have a constructive proof.

A solution to this problem is to decompose hypothetical provability into absolute provability and entailment. For absolute provability, the property that a sequent $\vdash A$ has a classical proof if and only if the sequent $\vdash \|A\|$ has a constructive one holds for all propositions, because atomic propositions have no proof. Thus, the sequent $H_1, ..., H_n \vdash A$ has a classical proof if and only if the sequent $\vdash \|H_1\| \Rightarrow^c ... \Rightarrow^c \|H_n\| \Rightarrow^c \|A\|$ has a constructive one. This leads to a system where we have only one notion of absolute provability, but two notions of entailment: "A has a proof from the hypothesis H" can either be understood as "$H \Rightarrow A$ has a proof" or "$H \Rightarrow^c A$ has a proof".

Definition 3 (Classical and constructive provability) *Classical provability is defined by the cut free sequent calculus rules of Figure 1. We say that the proposition A has a classical proof if the sequent $\vdash A$ does.*

Constructive provability, our main notion of provability, is obtained by restricting to sequents with at most one conclusion. This requires a

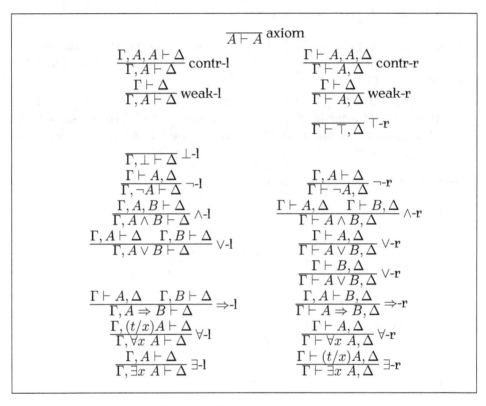

Figure 1: Sequent calculus

slight adaptation of the $\Rightarrow$-l rule

$$\frac{\Gamma \vdash A \quad \Gamma, B \vdash \Delta}{\Gamma, A \Rightarrow B \vdash \Delta} \Rightarrow\text{-}l$$

We say that the proposition A has a constructive proof if the sequent $\vdash A$ does.

Proposition 1 *If the proposition $\|A\|$ has a constructive proof, then the proposition A has a classical one.*

Proof. If the proposition $\|A\|$ has a constructive proof, then $\|A\|$ also has a classical proof. Hence, $\|A\|$ being classically equivalent to A, A also has a classical proof.

We now want to prove the converse: that if the proposition A has a classical proof, then $\|A\|$ has a constructive proof. To do so, we first introduce another translation where the top double negation is removed, when there is one.

Definition 4 (Light translation)

- $|P| = P$,

- $|\top| = \top$,

- $|\bot| = \bot$,

- $|\neg A| = \neg\neg\neg\|A\|$,

- $|A \wedge B| = (\neg\neg\|A\|) \wedge (\neg\neg\|B\|)$,

- $|A \vee B| = (\neg\neg\|A\|) \vee (\neg\neg\|B\|)$,

- $|A \Rightarrow B| = (\neg\neg\|A\|) \Rightarrow (\neg\neg\|B\|)$,

- $|\forall x\ A| = \forall x\ (\neg\neg\|A\|)$,

- $|\exists x\ A| = \exists x\ (\neg\neg\|A\|)$.

If $\Gamma = A_1, ..., A_n$ is a multiset of propositions, we write $|\Gamma|$ for the multiset $|A_1|, ..., |A_n|$ and $\neg|\Gamma|$ for the multiset $\neg|A_1|, ..., \neg|A_n|$.

Proposition 2 *If the proposition A is atomic, then $\|A\| = |A|$, otherwise $\|A\| = \neg\neg|A|$.*

Proof. By a case analysis on the form of the proposition A.

Proposition 3 *If the sequent* $\Gamma, |A| \vdash$ *has a constructive proof, then so does the sequent* $\Gamma, \|A\| \vdash$.

Proof. By Proposition 2, either $\|A\| = |A|$, or $\|A\| = \neg\neg|A|$. In the first case the result is obvious, in the second, we build a proof of $\Gamma, \|A\| \vdash$ with a $\neg$-l rule, a $\neg$-r rule, and the proof of $\Gamma, |A| \vdash$.

Proposition 4 *If the sequent* $\Gamma \vdash \Delta$ *has a classical proof, then the sequent* $|\Gamma|, \neg|\Delta| \vdash$ *has a constructive one.*

Proof. By induction on the structure of the classical proof of the sequent $\Gamma \vdash \Delta$. As all the cases are similar, we just give a few.

- If the last rule is the *axiom* rule, then $\Gamma = \Gamma', A$ and $\Delta = \Delta', A$, and the sequent $|\Gamma'|, |A|, \neg|A|, \neg|\Delta| \vdash$, that is $|\Gamma|, \neg|\Delta| \vdash$, has a constructive proof.

- If the last rule is the $\Rightarrow$-*l* rule, then $\Gamma = \Gamma', A \Rightarrow B$ and by induction hypothesis, the sequents $|\Gamma'|, \neg|A|, \neg|\Delta| \vdash$ and $|\Gamma'|, |B|, \neg|\Delta| \vdash$ have constructive proofs, hence the sequents $|\Gamma'|, \neg\|A\|, \neg|\Delta| \vdash$ and $|\Gamma'|, \|B\|, \neg|\Delta| \vdash$ have constructive proofs, thus, using Proposition 3, the sequent $|\Gamma'|, \neg\neg\|A\| \Rightarrow \neg\neg\|B\|, \neg|\Delta| \vdash$, that is $|\Gamma|, \neg|\Delta| \vdash$, has a constructive proof.

- If the last rule is the $\Rightarrow$-*r* rule, then $\Delta = \Delta', A \Rightarrow B$ and by induction hypothesis, the sequent $|\Gamma|, |A|, \neg|B|, \neg|\Delta'| \vdash$ has a constructive proof, thus, using Proposition 3, the sequent $|\Gamma|, \|A\|, \neg\|B\|, \neg|\Delta'| \vdash$ has a constructive proof, thus the sequent $|\Gamma|, \neg(\neg\neg\|A\| \Rightarrow \neg\neg\|B\|), \neg|\Delta'| \vdash$, that is $|\Gamma|, \neg|\Delta| \vdash$, has a constructive proof.

Proposition 5 *If the sequent* $\Gamma \vdash A$ *has a classical proof and* A *is not an atomic proposition, then the sequent* $\|\Gamma\| \vdash \|A\|$ *has a constructive one.*

Proof. By Proposition 4, as the sequent $\Gamma \vdash A$ has a classical proof, the sequent $|\Gamma|, \neg|A| \vdash$ has a constructive one. Thus, by Proposition 3, the sequent $\|\Gamma\|, \neg|A| \vdash$ has a constructive proof, and the sequent $\|\Gamma\| \vdash \neg\neg|A|$ also. By Proposition 2, as A is not atomic, $\|A\| = \neg\neg|A|$. Thus, the sequent $\|\Gamma\| \vdash \|A\|$ has a constructive proof.

Theorem 1 *The proposition* A *has a classical proof if and only if the proposition* $\|A\|$ *has a constructive one.*

Proof. By Proposition 1, if the proposition $\|A\|$ has a constructive proof, then the proposition A has a classical one. Conversely, we prove that if the proposition A has a classical proof, then the proposition $\|A\|$ has a constructive one. If A is atomic, the proposition A does not have a classical proof, otherwise, by Proposition 5, the proposition $\|A\|$ has a constructive proof.

Corollary 1 *The sequent $H_1, ..., H_n \vdash A$ has a classical proof if and only if the sequent $\vdash \|H_1\| \Rightarrow^c ...\|H_n\| \Rightarrow^c \|A\|$ has a constructive one.*

Proof. The sequent $H_1, ..., H_n \vdash A$ has a classical proof if and only if the sequent $\vdash H_1 \Rightarrow ...H_n \Rightarrow A$ has one and, by Theorem 1, if and only if the sequent $\vdash \|H_1\| \Rightarrow^c ...\|H_n\| \Rightarrow^c \|A\|$ has a constructive proof.

There is no equivalent of Theorem 1 if we add double negations after the connectors only. For instance, the proposition $P \vee \neg P$ has a classical proof, but the proposition $\neg\neg P \vee \neg\neg\neg\neg\neg\neg P$ has no constructive proof. O. Hermant (Hermant, 2012) has proved that there is also no equivalent of Theorem 1 if we add double negations before the connectors only. For instance, the proposition $(\forall x\ (P(x) \wedge Q)) \Rightarrow (\forall x\ P(x))$ has a classical proof, but the proposition $\neg\neg((\neg\neg\forall x\ \neg\neg(P(x) \wedge Q)) \Rightarrow (\neg\neg\forall x\ P(x)))$ has no constructive proof.

Let $H1, ..., H_n$ be an axiomatization of mathematics with a finite number of axioms, $H = H_1 \wedge ... \wedge H_n$ be their conjunction, and A be a proposition. If the proposition $H \Rightarrow A$ has a classical proof, then, by Theorem 1, the proposition $\|H\| \Rightarrow^c \|A\|$ has a constructive one. Thus, in general, not only the proposition A must be formulated with classical connectives and quantifiers, but the axioms of the theory and the entailment relation also.

Using Proposition 5, if A is not an atomic proposition, then the proposition $\|H\| \Rightarrow \|A\|$ has a constructive proof. In this case, the axioms of the theory must be formulated with classical connectives and quantifiers, but the entailment relation does not.

In many cases, however, even the proposition $H \Rightarrow \|A\|$ has a constructive proof. For instance, consider the theory formed with the axiom H "The union of two finite sets is finite"

$$\forall x \forall y\ (F(x) \Rightarrow F(y) \Rightarrow F(x \cup y))$$

—or, as the cut rule is admissible in sequent calculus, any theory where this proposition has a proof—and let A be the proposition "If the union of two sets is infinite then one of them is"

$$\forall a \forall b\ ((\neg F(a \cup b)) \Rightarrow (\neg F(a) \vee \neg F(b)))$$

which is, for instance, at the heart of the proof of Bolzano-Weierstrass theorem, then the proposition $H \Rightarrow \|A\|$ has a constructive proof

$$
\cfrac{
\cfrac{
\cfrac{
\cfrac{
\cfrac{
\cfrac{
\cfrac{
\cfrac{
\cfrac{
\cfrac{
\cfrac{
\cfrac{
\cfrac{
\cfrac{
\cfrac{
\cfrac{\overline{F(a), F(b) \vdash F(a)} \text{ axiom} \quad \cfrac{\overline{F(a), F(b) \vdash F(b)}\text{ axiom} \quad \overline{F(a \cup b), F(a), F(b) \vdash F(a \cup b)}\text{ axiom}}{F(b) \Rightarrow F(a \cup b), F(a), F(b) \vdash F(a \cup b)}\text{⇒-l}}{F(a) \Rightarrow F(b) \Rightarrow F(a \cup b), F(a), F(b) \vdash F(a \cup b)}\text{⇒-l}}{H, F(a), F(b) \vdash F(a \cup b)} \forall\text{-l}, \forall\text{-l}}{H, \neg\neg(\neg^c F(a \cup b)), F(a), F(b) \vdash} \text{¬-l, ¬-r, ¬-l, ¬-r, ¬-l, ¬-r, ¬-l}}{H, \neg\neg(\neg^c F(a \cup b)), F(a) \vdash \neg^c F(b)} \text{¬-r, ¬-l, ¬-r, ¬-l, ¬-r}}{H, \neg\neg(\neg^c F(a \cup b)), F(a) \vdash \neg^c F(a) \vee^c \neg^c F(b)} \text{¬-l, ¬-r, ∨-r, ¬-r, ¬-l}}{H, \neg\neg(\neg^c F(a \cup b)), \neg(\neg^c F(a) \vee^c \neg^c F(b)), F(a) \vdash} \text{¬-l}}{H, \neg\neg(\neg^c F(a \cup b)), \neg(\neg^c F(a) \vee^c \neg^c F(b)) \vdash \neg^c F(a)} \text{¬-r, ¬-l, ¬-r, ¬-l, ¬-r}}{H, \neg\neg(\neg^c F(a \cup b)), \neg(\neg^c F(a) \vee^c \neg^c F(b)) \vdash \neg^c F(a) \vee^c \neg^c F(b)} \text{¬-r, ¬-l, ∨-r, ¬-r, ¬-l}}{H, \neg\neg(\neg^c F(a \cup b)), \neg(\neg^c F(a) \vee^c \neg^c F(b)), \neg(\neg^c F(a) \vee^c \neg^c F(b)) \vdash} \text{¬-l}}{H, \neg\neg(\neg^c F(a \cup b)), \neg(\neg^c F(a) \vee^c \neg^c F(b)) \vdash} \text{contr-l}}{H, \neg\neg(\neg^c F(a \cup b)) \vdash \neg\neg(\neg^c F(a) \vee^c \neg^c F(b))} \text{¬-r}}{H \vdash (\neg^c F(a \cup b)) \Rightarrow^c (\neg^c F(a) \vee^c \neg^c F(b))} \text{¬-r, ¬-l, ⇒-r}}{H \vdash \forall^c a \forall^c b ((\neg^c F(a \cup b)) \Rightarrow^c (\neg^c F(a) \vee^c \neg^c F(b)))} \text{(¬-r, ¬-l, ∀-r, ¬-r, ¬-l)}^2}{\vdash H \Rightarrow \forall^c a \forall^c b ((\neg^c F(a \cup b)) \Rightarrow^c (\neg^c F(a) \vee^c \neg^c F(b)))} \text{⇒-r}
$$

In this case, even the proposition

$$
H \Rightarrow \forall a \forall b ((\neg F(a \cup b)) \Rightarrow (\neg F(a) \vee^c \neg F(b)))
$$

where the only classical connective is the disjunction, has a constructive proof.

Which mathematical results have a classical formulation that can be proved from the axioms of constructive set theory or constructive type theory and which require a classical formulation of these axioms and a classical notion of entailment remains to be investigated.

References

Allali, L. & Hermant, O. (2010), Cut elimination in classical sequent calculus modulo a super-consistent theory, Technical report.

Gentzen, G. (1969), *The Collected Works of Gerhard Gentzen*, chapter Über das Verhältnis zwischen intuitionistischer and klassischer Logik, pp. 53–67.

Girard, J.-Y. (1993), On the unity of logic, in 'Annals of Pure and Applied Logic', Vol. 59, pp. 201–217.

Gödel, K. (1932), 'Zum intuitionistischen aussagenkalkül', *Anzeiger der Akademie der Wissenschaften in Wien*.

Gödel, K. (1933), 'Zur intuitionistischen arithmetik und zahlentheorie', *Ergebnisse einses mathematischen Kolloquiums*.

Hermant, O. (2012), Personal communication.

Kolmogorov, A. (1925), 'On the principle of the excluded middle (russian)', *Matematicheskij Sbornik*.

www.ingramcontent.com/pod-product-compliance
Lightning Source LLC
LaVergne TN
LVHW012329060326
832902LV00011B/1785